ACOUSTIC EMISSION / MICROSEISMIC ACTIVITY

ACOUSTIC EMISSION / MICROSEISMIC ACTIVITY

Volume 1

Principles, Techniques, and Geotechnical Applications

H. Reginald Hardy, Jr.

The Pennsylvania State University, University Park, Pennsylvania, USA

CRC Press
Taylor & Francis Group
Boca Raton London New York

CRC Press is an imprint of the
Taylor & Francis Group, an **informa** business

Contents

Preface

In the late 1930s two researchers, L. Obert and W. I. Duvall, from the U.S. Bureau of Mines (USBM), carrying out sonic studies in a deep hard rock mine, discovered to their surprise that a stressed rock pillar appeared to emit micro-level sounds (Obert, 1977). Later, laboratory and field studies by these workers verified that this phenomenon, which is often referred to in nontechnical terms as "rock talk", is a measure of the mechanical stability of rock material and/ or the associated rock structure. In subsequent years, the early pioneer work of Obert and Duvall has led to the development of one of the truly unique techniques available in the geotechnical field. The understanding of "rock talk", or acoustic emission/microseismic (AE/MS) activity, as it will be referred to in this text, has developed considerably over the last 60 years. Today it is utilized as a routine tool in a number of geotechnical applications and new ones are rapidly becoming apparent. Through the efforts of an increasing number of researchers and practitioners in this field, improvement in the associated monitoring and analysis techniques, and the development of an improved appreciation of the basic processes involved, scientists and engineers are now better able to understand what stressed rock is "saying".

It is generally known today that many materials besides rock emit AE/MS activity when they are stressed and/or deformed. The phenomenon is in fact commonly observed in most solids including metals, ceramics, rocks, concrete, glass, wood, plastics, and even ice. Initial AE/MS research in the materials science area is generally assumed to have begun in the late 1940s with the work of Kaiser (1953) on metals, some ten years after the pioneering work of Obert and Duvall. Unfortunately, the use of different terms, i.e., acoustic emission and microseismic activity, to denote the phenomenon in the materials science and the mineral engineering fields seriously limited cross-fertilization until the late 1960s. As a result, for many years AE/MS research, and the associated technical publications in this field, have been widely scattered throughout various areas of engineering and science.

The lack of a unified compilation of the literature in the AE/MS field has been a serious disadvantage both to those actively involved as well as those interested students and potential practitioners anxious to learn more about the subject and its possible applications. More recently, however, as a result of the continuing efforts of Acoustic Emission Working Groups in the USA (AEWG), Europe and Japan; a series of Symposia sponsored by the American Society for Testing and Materials (ASTM), AEWG and the Society for Nondestructive Testing (ASNT), (e.g. Liptai et al., 1972; Spanner and McElroy, 1975; Hartman and McElroy, 1979; Drnevich and Gray, 1981; Ono, 1989 and Vahaviolos, 1991); the on-going International Acoustic Emission Symposia sponsored by the Japanese Society for Non-Destructive Inspection (JSNDI), (e.g. Kishi et al., 1994); the Journal of Acoustic Emission, now in its 16th year of publication, and a number a detailed publications (e.g. Drouillard, 1979; Matthews, 1983; McIntire, 1987; Spanner, 1974); workers in a wide range of disciplines are becoming better acquainted with the extensive literature available on the subject. Standards have also been developed by ASTM in a number of areas (Anon., 1996). These include characterization of instrumentation, calibration and mounting of sensors (transducers) as well as specific standards in a range of materials and special structures.

The following book is specifically oriented to the theory and application of AE/MS techniques in the geotechnical area. According to Roberts (1977) "Geotechnology is a field of study and engineering in which the applied geologist, the civil engineer and the mining engineer have a common interest". Therefore, in this text it is generally assumed that geotechnical applications include those in surface and underground mining, foundation engineering, tunneling, and rock and soil mechanics. In an attempt to unify efforts in the geotechnical area, the author, as Director of the Penn State Rock Mechanics Laboratory, in 1975 hosted the First Conference on Acoustic Emission/ Microseismic Activity in Geologic Structures and Materials. Subsequent conferences were held in 1978, 1981, 1985, 1991 and 1996. The proceedings of all six conferences have been published (Hardy, 1989, 1995 and 1998; Hardy and Leighton, 1977, 1980 and 1984) and these contain over 200 papers dealing with various basic and applied aspects of AE/MS activity in the geotechnical field, as well as an extensive bibliography.

The author has been intimately involved with the AE/MS area and its application in the geotechnical field now for some 30 years. During this time he has tried to keep in close contact with new developments, both in the USA and elsewhere, and to take an active part in promoting the future development and application of AE/MS techniques. Based on the experience gained through a variety of on-going AE/MS research projects at Penn State, the writer has in recent years published two relatively detailed reviews of the subject (Hardy, 1972, 1981). In the late 1970s he developed a graduate course on the subject while has subsequently been revised and updated to a two-semester course routinely presented within the geomechanics graduate program at Penn State. The present book, which is based on the earlier review papers, graduate course notes and current research, was developed as the text for this course. Where necessary the content has been updated, most recently in the spring of 2001.

Preparation of a text on the topic of geotechnical applications of AE/MS activity has presented a considerable challenge due to the strong interdisciplinary nature of the subject. For example, topics such as material behavior, stress wave propagation, electronic instrumentation, data acquisition and analysis, and signal processing must be included. Furthermore, consideration must be given to the practical applications of the AE/MS technique to field and laboratory problems in a wide range of geotechnical areas including those associated with geology, geophysics, and mining, petroleum, and civil engineering. It should be appreciated at the outset that there are always a number of inherent defects in any book covering such a specialized topic as AE/MS techniques. In particular, one can only cover the main aspects of the subject and even this coverage may be, intentionally or unintentionally, biased by one's specific interests and experience. Where possible the author has tried to keep the latter defect to a minimum.

Many of my colleagues, associates, and graduate students have contributed to the development of this book. In the early days the AE/MS research program at The Pennsylvania State University could never have been possible without the continuing support of Dean Charles Hosler and the late Dean Robert Stefanko. On my arrival as a faculty member in the Mining Department at Penn State in 1966, Professor Stefanko, head of the department at that time, assigned me an M.S. graduate student, Mr. Yoginda P. Chugh, presently a Professor at the University of Southern Illinois, as an advisee. It was clear that my knowledge of the associated research project, "acoustic emission during tensile loading of rock" was minimal and clearly was somewhat less than that of the student. However, we learned about AE/MS together, and a paper based on the student's research resulted in a project with the U.S. Bureau of Mines, and the initiation of a long term series of field and laboratory related AE/MS studies.

None of the AE/MS research would have been possible without the assistance of an outstanding Research Aide, and friend, Edward J. Kimble, Jr. For more than 25 years, Ed was personally responsible for the development of the majority of the facilities used in Penn State AE/MS laboratory and field studies, the planning and coordination of field studies, and subsequent organization of field data analysis. The author has also been fortunate to have had an excellent group of doctoral students working in the AE/MS area: notably Robert Belesky, Maochen Ge, A. Wahab Khair, Ran Young Kim, Gary L. Mowrey, Marek Mrugala, Xiaoqing Sun, and Erdal Unal.

I am also very grateful to the sponsors of our AE/MS research, in particular the National Science Foundation, the former U.S. Bureau of Mines, the American Gas Association, the Federal Aviation Authority, the Pennsylvania Department of Transportation, and IBM. In addition, university research funds were provided by Penn State's College of Earth and Mineral Sciences, the Department of Mineral Engineering, and the Pennsylvania Mining and Minerals Research Institute.

My wife, Margaret, contributed significantly to the preparation and publication of this book. Her patience, interest and encouragement over the years have been a major factor in my career. This book is dedicated to her.

H. Reginald Hardy, Jr.

University Park, Pennsylvania
July 2001

REFERENCES

Anon. 1996. *Annual Book of ASTM Standards*, Vol. 03.03, Nondestructive Testing, American Society for Testing and Materials, West Conshohocken, Pennsylvania.

Drnevich, V.P. & Gray, R.E. (eds.) 1981. *Acoustic Emissions in Geotechnical Engineering Practice*, STP 750, American Society for Testing and Materials, Philadelphia, 209 pp.

Drouillard, T.F. 1979. *Acoustic Emission: A Bibliography with Abstracts*, Plenum Publishing Company, New York, 787 pp.

Hardy, H.R., Jr. 1972. "Application of Acoustic Emission Techniques to Rock Mechanics Research", *Acoustic Emission*, Editors – R. G. Liptai, D. O. Harris and C. A. Tatro, STP 505, American Society for Testing and Materials, Philadelphia, pp 41-83.

Hardy, H.R., Jr. 1981. "Applications of Acoustic Emission Techniques to Rock and Rock Structures: A State-of-the-Art Review", *Acoustic Emissions in Geotechnical Engineering Practice*, STP 750, V. P. Drnevich and R. E. Gray, Editors, American Society for Testing and Materials, Philadelphia, pp. 4-92.

Hardy, H.R., Jr. (ed.) 1989. *Proceedings, Fourth Conference on Acoustic Emission/ Microseismic Activity in Geologic Structures and Materials*, The Pennsylvania State University, October 1985, Trans Tech Publications, Clausthal-Zellerfeld, Germany, 711 pp.

Hardy, H.R., Jr. (ed.) 1995. *Proceedings, Fifth Conference on Acoustic Emission/ Microseismic Activity in Geologic Structures and Materials*, The Pennsylvania State University, June 1991, Trans Tech Publications, Clausthal-Zellerfeld, Germany, 755 pp.

Hardy, H.R., Jr. (ed.) 1998. *Proceedings, Sixth Conference on Acoustic Emission/ Microseismic Activity in Geologic Structures and Materials*, The Pennsylvania State University, June 1996, Trans Tech Publications, Clausthal-Zellerfeld, Germany, 688 pp.

Hardy, H.R., Jr. & Leighton, F.W. (eds.) 1977. *Proceedings, First Conference on Acoustic Emission/Microseismic Activity in Geologic Structures and Materials*, The Pennsylvania State University, June 1975, Trans Tech Publications, Clausthal-Zellerfeld, Germany, 489 pp.

Hardy, H.R., Jr. & Leighton, F.W. (eds.) 1980. *Proceedings, Second Conference on Acoustic Emission/Microseismic Activity in Geologic Structures and Materials*, The

Pennsylvania State University, November 1978, Trans Tech Publications, Clausthal-Zellerfeld, Germany, 491 pp.

Hardy, H.R. Jr. & Leighton, F.W. (eds.) 1984. *Proceedings, Third Conference on Acoustic Emission/Microseismic Activity in Geologic Structures and Materials*, The Pennsylvania State University, October 1981, Trans Tech Publications, Clausthal-Zellerfeld, Germany, 814 pp.

Hartman, W.F. & McElroy, J.W. (eds.) 1979. *Acoustic Emission Monitoring of Pressurized Systems*, STP 697, American Society for Testing and Materials, Philadelphia, 225 pp.

Kaiser, J. 1953. "Untersuchungen uber das Auftreten Gerauschen Beim Zugversuch", Arkiv fur das Eisenhuttenwesen, Vol. 24, pp. 43-45.

Kishi, T., Mori, Y. & Enoki, M. (eds.) 1994. *Proceedings, 12th International Acoustic Emission Symposium*, Japanese Society for Non-Destructive Inspection, Tokyo, 636 pp.

Liptai, R.G., Harris, D.O. & Tatro, C.A. (eds.) 1972. *Acoustic Emission*, STP 505, American Society for Testing and Materials, Philadelphia, 337 pp.

Matthews, J.R. (ed.) 1983. *Acoustic Emission*, Gordon and Breach, Scientific Publishers, Inc., New York, 167 pp.

McIntire, P. 1987. *Nondestructive Testing Handbook, Vol. 5, Acoustic Emission Testing*, American Society for Nondestructive Testing, Columbus, Ohio, 603 pp.

Obert, L. 1977. "The Microseismic Method–Discovery and Early History", *Proceedings, First Conference on Acoustic Emission/Microseismic Activity in Geologic Structures and Materials*, The Pennsylvania State University, June 1975, Trans Tech Publications, Clausthal, Germany, pp. 11-12.

Ono, K. (ed.) 1989. 3rd World Meeting on Acoustic Emission, Charlotte, North Carolina, March 1989, *Extended paper summaries*, Journal of Acoustic Emission, Special Supplement, Los Angeles, California, 338 pp.

Roberts, A. 1977. *Geotechnology*, Pergamon Press, New York, p. 1.

Spanner, J.C. 1974. *Acoustic Emission: Techniques and Applications*, Intex Publishing Company, Evanston, Illinois, 274 pp.

Spanner, J.C. & McElroy, J.W. (eds.) 1975. *Monitoring Structural Integrity by Acoustic Emission*, STP 571, American Society for Testing and Materials, Philadelphia, 289 pp.

Vahaviolos, S.J. (ed.) 1991. 4th World Meeting on Acoustic Emission and 1st International Conference on Acoustic Emission and Manufacturing, Boston, September 1991, *Extended paper summaries and abstract*, American Society for Nondestructive Testing, Inc., Columbus, Ohio, 540 pp.

CHAPTER 1

Introduction

1.1 GENERAL

It is generally accepted today that most solids emit low-level seismic signals when they are stressed or deformed. A variety of terms, including acoustic emission, microseismic activity, seism-acoustic activity, subaudible noise, roof and rock talk, elastic shocks, elastic radiation, stress wave analysis technique (SWAT), and micro-earthquake activity are utilized by various disciplines to denote this phenomenon. It will be referred to throughout this text, however, as acoustic emission/microseismic (AE/MS) activity.

In recent years the application of AE/MS techniques in the general area of geotechnical engineering has rapidly increased. At present such techniques are in use, or under evaluation, for stability monitoring of underground structures such as mines, tunnels, natural gas and petroleum storage caverns, radioactive waste repositories, and geothermal reservoirs, as well as surface structures such as foundations, rock and soil slopes, bridge piers and abutments, and dams. The techniques, originally developed in an attempt to predict and reduce the incidence of rock bursts in hardrock, and later in coal mines, has now found extensive application in soils and soft rocks such as salt. Furthermore, besides their application in large scale field studies, AE/MS techniques are being successfully utilized in an increasing number of laboratory-scale studies both of a basic and an applied nature.

In essence the measurement of AE/MS activity in a field structure or laboratory specimen is relatively simple. A suitable transducer is attached to the structure or specimen, the output of the transducer is connected to a suitable monitoring system, and the acoustic signals occurring in the structure or specimen, due to internal or external stress or deformation, are suitably processed and recorded. It should be noted from the outset that the technique is an indirect one. The AE/MS technique does not directly determine basic mechanical parameters such as stress or strain, rather it determines the mechanical stability of a structure or specimen when it is subjected to stress or deformation.

1.2 BASIC CONCEPTS

Before proceeding to a detailed review of the subject, an outline of the fundamentals of the AE/MS technique will be presented, along with a brief discussion of typical AE/MS signals, measurement techniques, and a review of the historical development of the subject. A more comprehensive discussion of many of these factors is included later in the text.

1.2.1 Acoustic techniques

It is important to note that there are a number of so-called acoustic techniques presently in use to evaluate material behavior. Figure 1.1 illustrates two of these techniques commonly in use in the geotechnical area. The first, the sonic technique, utilizes two transducers; one, a transmitter, generates a mechanical signal within the material under study, the other, a receiver, monitors the transmitted signal and any modifications in it resulting from changes of stress and/or changes in the basic characteristics of the material. In contrast the AE/MS technique utilizes only a receiving transducer (one or more) which monitors self-generated (passive) acoustic signals ocurring within the material.

The sonic technique was employed by Obert and Duvall (1942) during early rock mechanics studies in a rock burst prone mine. Here a transmitter-receiver pair was attached to a stressed pillar in an attempt to measure changes in sonic velocity indicative of changes in pillar stress. When, during the course of their experiments, they removed the transmitter and still detected signals at the

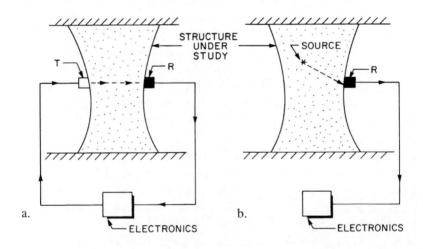

Figure 1.1. Two acoustic techniques utilized to evaluate the behavior of geologic structures. a. Sonic technique. b. AE/MS technique.

receiver, the AE/MS technique was born (Obert, 1977). Due to the variety of terms used to denote the AE/MS technique in the geotechnical area, particularly in Europe where the term seismo-acoustic activity is commonly used, it is important when reviewing the literature that the reader is sure which of the two acoustic techniques has actually been employed.

1.2.2   AE/MS sources

In geologic materials the origin of AE/MS activity is not well understood, but it appears to be related to processes of deformation and failure which are accompanied by a sudden release of strain energy. In such materials, which are basically polycrystalline in nature, AE/MS activity may originate at the micro-level as a result of dislocations, at the macro-level by twinning, grain boundary movement, or initiation and propagation of fractures through and between mineral grains, and at the mega-level by fracturing and failure of large areas of material or relative motion between structural units.

In contrast the sources of AE/MS activity in a variety of non-geologic materials have been investigated in considerable detail (Bassim, 1987; Wadley, 1987). In metallic materials plastic deformation is the primary source. In general, a number of sources with origins at the macroscopic level, such as crack growth, fatigue and stress corrosion cracking, and hydrogen embrittlement, have been recognized. At the microscopic level, sources associated with dislocations, microcracks, grain size effects, inclusions, coalescence of micro voids, and phase transitions such as Martensitic, and liquid-solid have been identified.

Clearly, further research to delineating the sources of AE/MS activity in geologic materials is of prime importance if a meaningful relationship between mechanical behavior and observed AE/MS activity is to be achieved.

1.2.3   AE/MS signals

It is assumed that the sudden release of stored elastic strain energy accompanying these processes generates an elastic stress wave which travels from the point of origin within the material to a boundary, where, using a suitable transducer, it is observed as an AE/MS signal or a discrete AE/MS event. Chapter 2 considers the propagation and detection of AE/MS signals in considerable detail.

The fundamental frequency character of an observed AE/MS signal depends on the characteristics of the source, and the distance between the source and the monitoring transducer. Frequencies below 1 Hz have been observed at large scale field sites, whereas in laboratory studies AE/MS signals have often been observed to contain frequencies greater than 500 kHz.

Figure 1.2 indicates the frequency range over which AE/MS and other associated studies have been conducted, and Figure 1.3 illustrates a number of typical AE/MS events recorded during various field and laboratory studies. It should also be noted that most AE/MS signals contain both a compressional (P-wave) and shear (S-wave) component. In some cases these are difficult to distinguish; however, the AE/MS signal presented in Figure 1.3b clearly shows

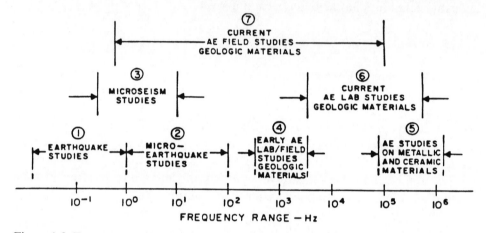

Figure 1.2. Frequency range over which AE/MS and other associated studies have been conducted.

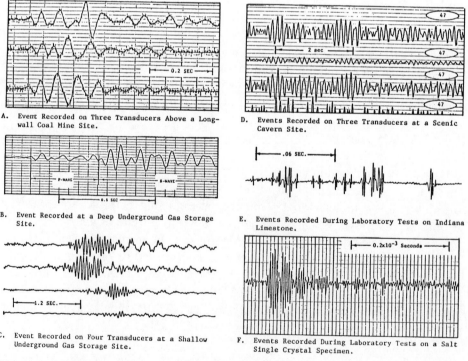

A. Event Recorded on Three Transducers Above a Long-wall Coal Mine Site.

B. Event Recorded at a Deep Underground Gas Storage Site.

C. Event Recorded on Four Transducers at a Shallow Underground Gas Storage Site.

D. Events Recorded on Three Transducers at a Scenic Cavern Site.

E. Events Recorded During Laboratory Tests on Indiana Limestone.

F. Events Recorded During Laboratory Tests on a Salt Single Crystal Specimen.

Figure 1.3. Typical AE/MS events recorded in various field and laboratory studies undertaken by the Penn State Rock Mechanics Laboratory.

both components. A third component, due to surface waves, is also observed in situations where the monitoring transducer is located on or near a free surface.

Frequency analysis of an individual AE/MS event indicates that it contains a spectrum of different frequencies. The form of the observed spectrum is a result

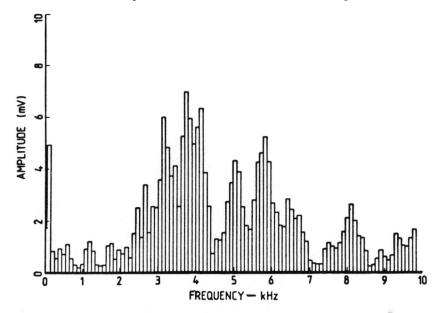

a.

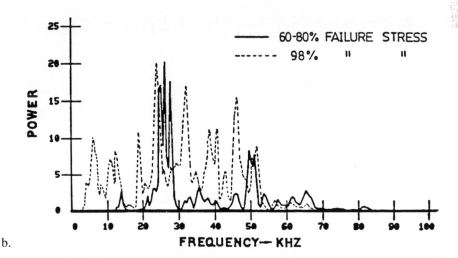

b.

Figure 1.4. Typical frequency spectra for AE/MS events monitored in coal under field and laboratory conditions. a. Underground coal mine event (after McKavanagh and Enever, 1980). b. Laboratory event during test on bituminous coal under unconfined compressive stress (after McCabe, 1980).

of two separate factors, namely: the spectrum of the event at its source, and modifications incurred during its propagation from the source location to the point of observation (transducer location). For example, Figure 1.4 illustrates typical frequency spectra for AE/MS events monitored in coal under both field and laboratory conditions. It is clearly shown that much higher frequencies are observed under laboratory conditions. Attenuation, which in geologic materials is often highly frequency dependent, plays a major role in modifying the original AE/MS source spectrum. The ability to detect AE/MS events at a distance from their source is thus dependent on the source spectrum, the degree of frequency dependence of the attenuation, the distance of the transducer from the source, and the band-width and sensitivity of the transducer and the associated monitoring system.

As a general rule, attenuation increases with frequency, thus at large distances from a source only the low frequency components of the event will be observed. Furthermore, if the source spectrum of the event contains no significant low frequency components there will be a critical distance, or range, beyond which the event cannot be detected. In other words, there is a relationship between range and frequency of the form shown in Figure 1.5.

Normally the range vs frequency curve, the form of which depends on the characteristics of both the associated geologic material and the specific monitoring system utilized, will be more complex than that shown. Such range versus frequency data is extremely important for the design of an optimum monitoring arrangement. For example, if the AE/MS source spectrum was

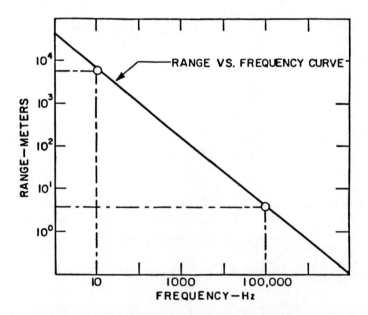

Figure 1.5. Typical range vs frequency data for AE/MS signals.

assumed to be wide, for example, containing components from 10 Hz to 100 kHz, the maximum range of the system would be approximately 3000 meters. The data shown in Figure 1.5 is also useful if it is desired to restrict the monitoring range of the system. For example, for the same source spectrum, by using a transducer only sensitive to signals of 100 kHz or higher, or by introduction of a suitable electronic filter in the system which would eliminate signals below 100 kHz, the range of AE/MS detection would be limited to approximately five meters or less.

1.2.4 Monitoring techniques

The techniques used to monitor AE/MS activity in the field and the laboratory are essentially the same, although the scale, and as a result, the frequency characteristics of the monitoring system will usually be somewhat different. A detailed discussion of various types of monitoring systems will be presented later in Chapters 3 and 4. At this stage a simple example of a field monitoring system will serve to introduce the reader to AE/MS monitoring techniques.

In the field, AE/MS signals are obtained by installing a suitable transducer, or usually a number of transducers (an array), in locations where they can detect any activity which may be originating in the structure under study. The overall monitoring system involves a transducer, an amplifying and filtering system, and a recorder. Figure 1.6 illustrates a typical system which might be used, for example, to monitor the AE/MS activity occurring in a shallow underground mine. In such an application the transducer could be a geophone (velocity gage) located in a suitable borehole perhaps 8-30 m below surface. In this case the transducer would monitor changes in ground velocity caused by the presence of AE/MS activity.

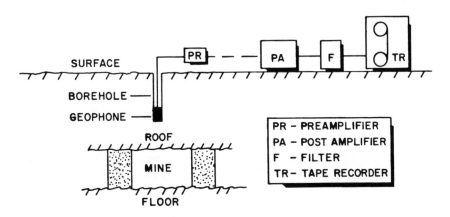

Figure 1.6. Block diagram of a typical AE/MS field monitoring system.

The output of the transducer, resulting from the presence of typical AE/MS activity, is normally only of the order of microvolts, and it is necessary to amplify this small signal, without introducing noise and distortion, to a sufficient level in order to record it. Furthermore, to properly match the impedance of the transducer to the relatively low input impedance of the post-amplifier, it is necessary to employ an intermediate preamplifier. A band-pass filter is also normally included in such systems in order to eliminate undesirable extraneous low and high frequency signals. Past experience has shown that a magnetic tape recorder often provides the most satisfactory technique for recording AE/MS signals, since it introduces a minimum of distortion, allows for an extended monitoring period, and makes it possible to conveniently analyze the recorded data at a later date.

AE/MS monitoring facilities for use in laboratory studies often tend to be more sophisticated than those employed in field applications. Although in some cases systems similar to that shown in Figure 1.6 have been employed, many of the current laboratory systems do not record the actual AE/MS signals themselves. In contrast they are designed to process the AE/MS signals and provide such parameters as total event count, event rate, etc. Such parameters may be recorded continuously during on-going laboratory tests along with such data as applied load, specimen strain, etc.

1.2.5 Source location

One of the major advantages of the AE/MS technique over other geotechnical monitoring techniques is its ability to delineate the area of instability. From a fundamental point-of-view, accurate AE/MS source location in a specimen or a structure is extremely important. First, unless the actual source location is accurately located, it is impossible to estimate the true magnitude of an observed AE/MS event. That is, a series of small observed events may be due to a weak source located close to the AE/MS transducer or due to a strong source located a considerable distance away. Secondly, in order to determine the mechanism responsible for the observed activity, it is necessary that the location of the source be accurately known. In general, source location techniques involve the use of a number of monitoring transducers located at various points throughout the body (specimen or structure) under study. Such a set of transducers is termed an "array". Chapter 5 discusses source location techniques in further detail.

Figure 1.7 illustrates, for example, a typical field situation where AE/MS techniques are being employed to monitor the mechanical stability of an underground storage cavern. Here suitable transducers have been installed at accurately known locations, and data from these is monitored during product input, output, or long-term storage. AE/MS activity occurring during such an evaluation is detected at each transducer at a different time depending on

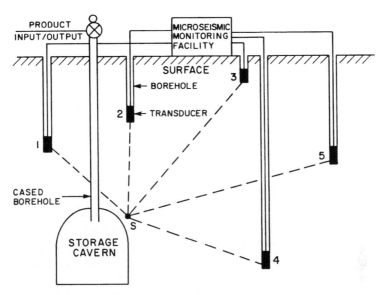

Figure 1.7. Transducer array installed to monitor the stability of an underground storage cavern, and to locate the source(s) of the instability.

the distance between the particular transducer and the AE/MS source. The difference in arrival-time between the closest transducer and each of the others yields a set of arrival-time-differences which, along with the geometry of the transducer array and the velocity of propagation in the material, may be used to determine the spatial coordinates of the AE/MS source. In such a study, source location provides the advantage of being able to determine the location of any instabilities which may influence the safety and operating performance of the facility and, furthermore, provides a means of eliminating, from subsequent analysis, AE/MS activity occurring outside the immediate area of the facility. This data processing technique is known as spatial filtering.

1.2.6 Review of basic concepts

In the preceding section a number of the basic concepts associated with the AE/MS technique, as it applies to both laboratory and field studies in the geotechnical area, have been briefly discussed. In review the important AE/MS concepts are as follows:
1. The AE/MS technique is a passive, indirect technique.
2. AE/MS activity originates as an elastic stress wave at locations where the material is mechanically unstable.
3. The associated stress wave propagates through the surrounding material undergoing attenuation as it moves away from the source.
4. With suitable instrumentation, AE/MS activity may be detected at locations a considerable distance from its source.

5. The useable spatial range of the technique is dependent on the frequency content of the source and the characteristics of the media and the monitoring facility.
6. The character of the observed AE/MS signals provide indirect evidence of the type and degree of the associated instability.
7. Analysis of data obtained from a number of transducers (array) make it possible to determine the actual location (i.e., spatial coordinates) of the source.

1.3   AE/MS PARAMETERS AND DATA ANALYSIS

A variety of parameters are utilized to describe AE/MS signals observed during laboratory and field studies. In many of the laboratory studies the AE/MS signals themselves are not normally recorded, due to their high frequency characteristics (e.g. f > 50 kHz). Instead the signals are processed on-line using analog or digital techniques to provide various related parameters such as accumulated activity or event rate. In large scale field studies, where the frequency content is normally relative low (e.g. f < 1000 Hz), a similar approach is often taken; however, in many cases the complete AE/MS signals are recorded for later analysis.

In general, the analysis of AE/MS signals presents a number of difficulties, namely:
1. Although in laboratory studies a single channel of data (i.e., a single transducer) is often sufficient, in most field cases it is necessary to deal simultaneously with at least five channels of data, often seven or more.
2. AE/MS events occur randomly in both time and amplitude.
3. The frequency content of AE/MS signals from within a specimen or structure may vary considerably from point to point. This is particularly true in field studies.
4. Separation of AE/MS signals from the ambient or background noise in some cases may be difficult, if not impossible.

At this point, a brief review of the more important AE/MS parameters will be presented. A detailed discussion of the procedures utilized for the storage and processing of AE/MS signals and related parameters will be included later in this book.

1.3.1   Important AE/MS parameters

In general, AE/MS signals, as illustrated in Figure 1.8, are randomly occurring transients whose characteristics depend on the mechanical properties of the laboratory specimen or field structure under study, and on the degree and type of instability involved. Although such signals have until recently been described

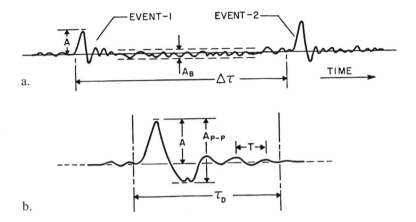

Figure 1.8. Typical section of AE/MS data and an expanded AE/MS event. a. Typical section of microseismic data. b. Expanded event.

mainly in terms of such factors as event rate and source location, they do contain considerable additional information making it possible to characterize them in terms of a number of other useful parameters. Such parameters may be determined for single- or multi-channel data. In the latter case more than one monitoring transducer is in use and data from each transducer is acquired and processed on a separate data channel.

Figure 1.8 illustrates diagrammatically a section of typical AE/MS data along with an expanded version of one of the events. In the time domain such data is most commonly described in terms of the following parameters:

1. *Accumulated Activity* (N): The total number of events observed during a specific period of time.
2. *Event Rate* (NR): The number of events (ΔN) observed per unit time (Δt).
3. *Amplitude* (A): The peak (maximum) value of each recorded event.
4. *Peak-to-Peak* Amplitude (A_{p-p}): The amplitude measured between the maximum positive and negative peaks of the event.
5. *Background Amplitude* (A_B): The signal level present in the absence of well-defined events. A_B as shown in Figure 1.8 is actually the peak-to-peak background amplitude.
6. *Signal + Noise-to-Noise Ratio* (SNR): The ratio of the event plus background amplitude to the background amplitude.
7. *Energy* (E): The square of the event amplitude (A).
8. *Accumulated Energy* (ΣE): The sum of the energy emitted by all events observed during a specific period of time.
9. *Energy Rate* (ER): The sum of the energy emitted by all events observed per unit time (Δt).

10. *Period* (T): The time between successive peaks of the event.
11. *Average Fundamental Frequency* (f): The reciprocal of the average period computed over n cycles of the event.
12. *Event Duration* (τ_D): The total time of occurrence for an individual event.
13. *Time-Between-Events* $(\Delta\tau)$: The time between successive events, i.e., the time between the beginning of one event and the beginning of the next.

1.3.2   Frequency spectra

AE/MS events (and AE/MS signals in general) may also be described in terms of their frequency spectra. In general, any transient signal, such as an AE/MS event, may be considered to be the superposition (i.e., algebraic sum) of a number of sinusoidal signals of specific frequency and amplitude. It is possible therefore to represent such a signal in either the time domain or the frequency domain. Mathematically, the conversion between these two domains may be carried out using the Fourier integral which has the general form:

$$G(t) = \frac{1}{\pi}\int_o^\infty S(\omega)\cos[\omega t+\phi(w)]d\omega \qquad \text{(Eq. 1.1)}$$

where $\omega = 2\pi f$; G(t) and $S(\omega)$ represent, respectively, the amplitude of the signal in the time and frequency domain; f is the frequency; t is the time; and $\phi(\omega)$ is the phase factor. Usually the function G(t) is referred to as the "waveform" of the signal and $S(\omega)$ or alternately S(f) is referred to as the "frequency spectrum" of the signal. $S(\omega)$ or S(f) are normally obtained from G(t) using some form of hard-wired frequency spectrum analyzer, or a suitable software program.

Figure 1.9 for example, illustrates the signal associated with a typical underground AE/MS event and its associated frequency spectra. The spectra indicates that frequency components at least as high as 1.5 kHz are present; however, the major components lie in the range 200-500 Hz. Further details in regard to the subject of frequency spectral analysis are available elsewhere (e.g., Bath, 1974; Blackman and Tukey, 1958; and Koopmans, 1974; and Robinson and Dussani, 1986).

1.3.3   Source location coordinates

As noted earlier, one of the major advantages of the AE/MS technique is that, by analysis of data from a suitable transducer array, it is possible to locate the source of the observed instability, i.e., to determine the source location coordinates. An analytical technique, generally known as the travel-time-difference method, is often employed for source location analysis. Considering the source location problem in more general terms, Figure 1.10 illustrates a

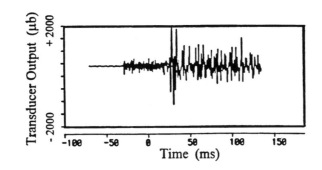

a.

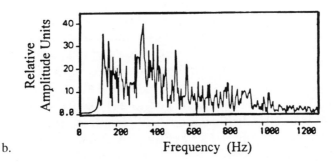

b.

Figure 1.9. AE/MS event signal and associated frequency spectra (after Baria et al., 1989). a. AE/MS event signal. b. Frequency spectra.

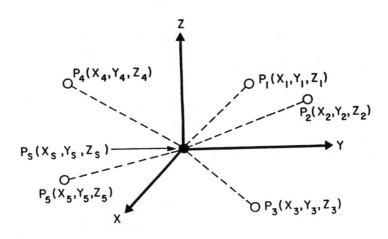

Figure 1.10. Geometry of a typical AE/MS transducer array. (Transducers are located at points $P_1,...,P_5$ with the AE/MS source at point P_s).

typical transducer array containing five transducers surrounding the AE/MS source, which is assumed to be located at point P_s.

Computerized methods for determining source locations often utilize a least-squares iteration technique which searches for the best average solution for a set of equations which contain the transducer coordinates, X_i, Y_i and Z_i ($i = 1,2,3,...,n$), and the corresponding AE/MS arrival-times and velocities, t_i and V_i. The method involves the solution of the following set of i equations:

$$(X_i - X_s)^2 + (Y_i - Y_s)^2 + (Z_i - Z_s)^2 = V_i^2(t_i - T_s)^2 \qquad (Eq.1.2)$$

where X_s, Y_s, and Z_s are the true coordinates of the source, T_s is the true origin time, and the other parameters have been defined earlier. As there are four unknowns, X_s, Y_s, Z_s, and T_s, at least four equations (and therefore, data from four transducer stations) are necessary to solve the equations exactly. However, due to inherent experimental errors related to each equation (e.g., errors in transducer locations, determination of arrival-time differences, and velocities), at least five equations (data from five transducer stations) should be employed in order that the set of equations defined by Equation 1.2 are over-determined. This allows the computer program to minimize such error terms and to arrive at a best fit or average solution for the equation set. Additional equations (> 5) will usually yield an improved value for the average solution because of further redundancy in the associated equations.

Accurate determination of source location coordinates involve a number of factors, the most critical being the quality of the velocity data. A detailed discussion of AE/MS source location methods is presented later in Chapter 5.

1.4 HISTORICAL DEVELOPMENT

Prior to considering in detail the AE/MS basic concepts, the techniques presently in use for monitoring and analysis of AE/MS data, and a number of current applications, a brief review of the historical development of the AE/MS technique in the geotechnical area will be presented. Further details in this regard are available in papers by Hardy (1972, 1981), Koerner et al. (1981), and Lord and Koerner (1979), and in the proceedings of the on-going Penn State Conference series (Hardy, 1989, 1995, 1998; Hardy and Leighton, 1977, 1980, 1984). A detailed history of the general subject of acoustic emission by Drouillard (1996) is available along with a comprehensive bibliography (Drouillard, 1979) which includes approximately 200 entries as well as a historical review of the general development of the subject.

1.4.1   Interdisciplinary nature of AE/MS development

There is little doubt that the initial development of the AE/MS technique
was associated with studies in the geotechnical field. In this regard the
pioneering work of Obert and Duvall (1942) in the late 1930s, appears to
represent the true beginning of the subject. Although in subsequent years,
AE/MS studies in mining and other geotechnical areas continued to develop,
many of the major developments, however, occurred within the disciplines of
geophysics and material science. Unfortunately, due to the wide differences
in the general interests and the materials under study in the different areas,
and the development of separate and somewhat unique nomenclature, for a
number of years the interaction between the various disciplines involved with
AE/MS research has been relatively limited. In the last 25 years, due in part
to a resurgence of interest in the AE/MS method by those in the geotechnical
field, it has become increasingly evident that a common field of interest exists.
Furthermore it is clear that studies in the various disciplines in most cases
mainly differ in respect to the scale of the specific problem under investigation,
and the frequency range of the associated AE/MS activity.

Figure 1.2, presented earlier, illustrates the wide frequency range
(approximately $10^{-2} - 10^6$ Hz), over which AE/MS techniques have been utilized
to evaluate structural and material stability. In general, studies associated with
the very low frequencies ($f < 10^2$ Hz) lie in the general domain of seismology,
whereas those at the very high frequencies ($f > 10^5$ Hz) are for the most
part associated with material science, including fields such as metallurgy,
ceramics, polymer science, etc. In contrast the frequencies of interest for field
and laboratory studies in the general geotechnical area (i.e., rock and soil
mechanics) extend over a wide central region of the overall frequency range
($10^0 < f < 5 \times 10^5$ Hz).

It is clear, therefore, that AE/MS studies in the geotechnical field overlap
at low frequencies with seismology, and at high frequencies with material
science related AE/MS studies. This presents a number of advantages since
specialized instrumentation and experimentation techniques have been under
development for some years in both areas. Furthermore, in the lower frequency
range there is an increasing interest by the seismologist in earthquake and
microearthquake activity associated with various structural instabilities
occurring during geotechnical activities such as mining, underground storage,
reservoir impounding, etc. This interest has, in recent years, developed to
the extent that it now represents a specialized area of seismology, namely,
"induced seismicity" (e.g. Milne, 1976). In the geotechnical area the majority
of AE/MS studies have been directly or indirectly related to the investigation
of various practical field problems. Clearly, since the beginning, field related
problems have been the major driving force in the development of the AE/MS
technique.

Table 1.1 briefly outlines a number of the major developments associated with the field-oriented applications of the AE/MS technique during the last 60 years. The following sections will present a brief historical review of the development of the AE/MS technique, including major developments both in the laboratory and in the field.

1.4.2   Early studies

As indicated earlier, initial AE/MS studies associated with geological materials were mainly undertaken to study the stability of underground mining operations. During the late 1930s and early 1940s, Obert and Duvall (1942) showed that

Table 1.1. Historical review of major geotechnical field-oriented AE/MS studies.

Date[1]	Investigators	Study Area
Early Studies		
1939	Obert and Duvall (USA)	Hardrock mines
1940[2]	USA/Canada/USSR	Various field applications
1953	Buchheim (Germany)	Coal and salt mine studies
1955	Crandel (USA)	Tunnel safety
1959	Vinogradov (USSR)	Mine studies
1962	Beard (USA)	Tunnel safety
Transitional Studies		
1963	Cook (South Africa)	Rock bursts
1964	Susuki et al. (Japan)	Mine studies
1965	Neyman et al. (Poland)	Coal mine safety
1966	Scholz (USA)	Source location techniques
	Mogi (Japan)	
1967	Goodman and Blake (USA)	Slope stability
1969	Duvall, Blake and Leighton (USA)	Rock bursts
1970	Stas et al. (Czechoslovakia)	Rock bursts
1970	Baule et al. (Germany)	Rock bursts
Recent Studies		
1971	Hardy (USA)	Coal mine safety
		Storage reservoir stability
1972	Koerner et al. (USA)	Soil studies
1975	Leighton (USA)	Coal mine safety
1976	Hayashi (Japan)	In-situ stress evaluation
1976	Albright (USA)	Hydraulic fracturing of geothermal reservoirs
1978	Will (Germany)	Coal mine safety
1978	McKavanagh and Enever (Australia)	Coal outbursts
1978	Van Zyl Brink (South Africa)	Rock bursts
1981	Flach and Hente (Germany)	Radioactive waste repository stability

Tabel 1.1 continued:

Date[1]	Investigators	Study Area
Recent Studies (continued)		
1981	Albright and Pearson (USA)	Salt cavern stability
	Gehl and Thoms	
1981	Frantti (USA)	Mine subsidence
1982	Ohtsu (Japan)	Concrete
1982	Niitsuma (Japan)	Hydraulic fracturing of geothermal reservoirs
1983	Holcomb (USA)	Kaiser Effect
	Hardy and Zhang (USA)	
1983	Baria et al. (UK)	Hydraulic fracturing of geothermal reservoirs
1984	Hardy (USA)	Sinkhole Monitoring
		Rock Slope Stability
1984	Young (Canada)	Rock bursts
1984	Wood and Harris (Australia)	Massive structures
1985	Carabelli (Italy)	Gravity dam stability
1985	Armstrong (USA)	Earthquake prediction
1985	Ge and Hardy (USA)	Transducer array geometry
1988	Taioli and Hardy	Acoustic waveguides
1990	Ge (Canada)	Source location optimization
1990	Niitsuma et al. (Japan)	Source location using spectral matrix analysis
1993	Niitsuma et al. (Japan)	Crack waves
		Triaxial doublet analysis
1995	Itakura et al. (Japan)	Fractural analysis

[1] Dates listed are approximate only, in some cases they represent dates of initiation of research, in other cases dates of major publications. Specific research studies will be referenced separately through this book.
[2] During the period 1940-1960s considerable government supported research was underway.

the AE/MS rate increased as laboratory specimens and field structures became more highly loaded. Conversely, as equilibrium was reached after a major structural failure, the rate decreased. In other words, the AE/MS rate appeared to be a factor indicative of the degree of instability of the specimen or the field structure. With the exception of a few other isolated basic studies, this early work of Obert and Duvall provided the basis for the majority of later AE/MS studies carried out in North America, and had a strong influence on studies carried out in Europe and elsewhere.

During this early period government agencies, both in the United States and Canada, became involved in studies related to underground mining. These studies were initiated as a result of difficulties experienced in mining at increasing depths or in highly stressed zones, the most spectacular of these being the sudden violent failure of mine structures known as rock bursts. It is interesting to note that following the work of Obert and Duvall many of the rock mechanics studies undertaken by the U.S. Bureau of Mines (USBM) included AE/MS monitoring as part of the associated monitoring program (e.g., Merrill, 1954, 1957; Merrill and Morgan, 1958).

Considerable application of AE/MS techniques to underground mining was also underway in Europe and Asia during this period. Although details of these studies are not generally available in English, a monograph by Antsyferov (1966), an English translation of which is available, does provide an interesting insight into the AE/MS activities underway in Russia during this period; and an early paper by Buchheim (1958) discusses the use of AE/MS techniques in European coal and salt mines. Aside from the mining area, some of the early AE/MS field studies were carried out in relation to tunnel development projects. For example, early papers by Crandell (1955) and Beard (1962) discuss the use of AE/MS techniques for evaluation of tunnel safety during development work.

In general, AE/MS laboratory research was relatively inactive during this period. Although the USBM carried out a limited number of studies (e.g., Obert and Duvall, 1945), the majority of the laboratory related work was undertaken by researchers interested in the application of AE/MS techniques to seismology and earthquake prediction (e.g., Ishimoto and Iida, 1939, Suzuki, 1953, 1954).

1.4.3   Transition period studies

There is little doubt that the 1960s was an important transition period in the development of AE/MS techniques. During the period 1940-1960, AE/MS studies were carried out during numerous underground research programs but in many cases little real use of the resulting data appeared to have been made. In the early 1960s a renewed and more detailed interest in the field applications of AE/MS techniques developed, and by the late 1960s a number of important studies were underway.

During this period more sophisticated techniques for monitoring underground AE/MS activity were investigated, and in particular, techniques for accurate source location were developed. During the early 1960s the work of Cook (1963), who developed a refined monitoring system for use in the South African Witwatersrand gold mining area, was of major importance. In his studies geophones were used as AE/MS transducers and monitoring for the most part was limited to a relatively low frequency range (15-300 Hz). AE/MS studies have continued in South Africa and the more recent activities will be discussed later.

In the 1960s AE/MS research at the USBM accelerated rapidly. In contrast to the studies in South Africa, the USBM studies involved AE/MS monitoring over a much wider frequency range, commonly 20 Hz to 10 kHz. In these studies Blake and Leighton (1970) utilized commercially available piezo-electric accelerometers as AE/MS transducers and studies were carried out in a number of hard rock mines.

During the transition period, general research in the field of slope stability, associated with both open pit mining and various civil engineering projects, increased rapidly; and AE/MS techniques were investigated by a number of workers as a tool for monitoring slope stability (e.g., Paulsen et al., 1967; Broadbent and Armstrong, 1969; Wisecarver et al., 1969). In such studies automatic long-term monitoring systems are desirable and a number of special systems were developed for this purpose.

Laboratory research in the AE/MS area also increased substantially during this period. Studies by a number of workers (Barron, 1970; Mae and Nakao, 1968; Mogi, 1962; Perani and Thénoz, 1969; Scholz, 1968A; Suzuki et al., 1964) confirmed the fact that under both uniaxial and triaxial compressive stress the AE/MS rate increases rapidly as the compressive failure stress is reached. Goodman (1963) studied the effect of repeated uniaxial compressive loading and unloading on AE/MS activity. It was observed that if the load was not increased beyond a critical level, referred to as the point of "accelerated rock noise activity", then fewer events were detected during the second and subsequent loading cycles. If loads higher than the critical level were applied, however, the rate of noise emission was found to increase in subsequent cycles. These studies later lead to the study of the so-called "Kaiser effect" as a means of evaluating in-situ stress using tests on field core specimens. Brown (1965) and Brown and Singh (1966) studied the behavior of a number of rock types under incremental tensile stress. From the data obtained in these experiments the authors postulated that AE/MS energy was the most suitable factor for describing the observed behavior.

During the period, a number of workers (e.g., Chugh et al., 1972; Suzuki et al., 1964) considered the frequency characteristics of individual AE/MS events with the aim of determining a relationship between frequency content and other physical properties. In general these studies indicated that many difficulties are involved in the study of AE/MS frequency spectra. These include the effect of inherent resonant frequencies in the test specimen and loading system, the frequency dependent attenuation characteristics of the test material, and the necessity of insuring that the overall monitoring system, including the transducer, has a flat frequency response in the region of interest.

Studies to evaluate the relationship between AE/MS activity and inelastic behavior were also carried out for both uniaxial and triaxial loading conditions by a number of workers. Brown (1965) and Brown and Singh (1966) carried out experiments under uniaxial tensile stress. Similar studies under uniaxial stress were carried out by Hardy et al. (1970) which suggested a nearly linear correlation existed between creep strain and AE/MS activity for a variety of rock types. Scholz (1968a,b) studied the relationship between microcracking, as evidenced by observed acoustic emission, and inelastic behavior for a number of rock types under both uniaxial and triaxial stress. Based on these studies he developed a theoretical relationship between microcracking frequency and applied stress. In a later paper Scholz (1968c) developed a mathematical relationship between observed creep and microfracturing. Whether creep is due to microfracturing or vice versa (Savage and Mohanty, 1969) is still debatable; however, it is certain that in geologic materials these two phenomenon are strongly related.

As noted earlier in this chapter, accurate source location in both laboratory specimens and field structures is extremely important. During the transition period pioneering work in this area, on the laboratory scale, was carried out by both Scholz (1968a,b) and Mogi (1968).

1.4.4   Recent studies

By the early 1970s a number of important AE/MS concepts had been developed, field and laboratory techniques and instrumentation were rapidly improving and a variety of AE/MS studies were underway. During the 1970s field-oriented AE/MS research continued to expand at the USBM and extensive studies in both coal and hard rock mines were carried out. The studies continued into the early 1990s. Research also continued in the hard rock mines of South Africa; AE/MS studies were initiated at hard rock mines in Australia, Sweden, Canada and Australia; and detailed coal mine studies were underway in Poland, Czechoslovakia, Germany and Japan.

During the early 1980s, studies were reported on a number of new field related areas. These included the application of AE/MS techniques to a variety of soil mechanics problems (Koerner et al., 1981); evaluation of the stability of underground gas storage reservoirs (Hardy, 1980), salt caverns (Albright et al., 1984), and radioactive waste disposal facilities (Hente et al., 1984; Majer et al., 1984); monitoring of geothermal masses during hydrofracturing and energy utilization (Anon, 1978; Batchelor et al., 1983; Niitsuma et al., 1984); and the development of an extensive Canadian effort in the study of rock bursts in underground hardrock mines (Young et al., 1987). Studies later in the 1980s included the AE/MS monitoring of massive structures (Wood et al., 1984, Carabelli, 1989); earthquake prediction using AE/MS data (Armstrong and Stierman, 1989; Dunegan, 1998), and investigation of the role of transducer array geometry in source location accuracy (Ge and Hardy, 1988).

During the 1990s many of the studies initiated earlier continued. Important new research included revolutionary improvements in mine source location techniques (Ge and Mottahed, 1993), and the ever increasing sophistication in AE/MS studies of geothermal reservoirs involving crack waves, and spectral matrix and triaxial doublet analysis undertaken by Professor Niitsuma and his associates at Tohoku University, Sendai, Japan (Asanuma and Niitsuma, 1995; Nagano et al., 1995, 1998).

Finally, it should be noted that since the early 1970s an increasing number of laboratory oriented AE/MS studies have also been underway, and such studies have accelerated during the late 1980s. These include detection of failure in pressurized model caverns (Khair, 1972); fundamental studies of the behavior of salt, including dissolution, yield point and creep characteristics (Hardy, 1996; Hardy et al., 1980; Meister, 1980; Stead and Szczepanik, 1995); evaluation of rock bolt anchor stability (Ballivy et al., 1998; Unal and Hardy, 1984); fundamental studies aimed at correlating stress induced microscopic changes in rock fabric with observed deformation and AE/MS activity (Fonseka et al., 1975; Montoto et al., 1984); and the behavior of concrete (Ohtsu, 1994). Other recent studies include basic hydrofracturing phenomena (Byerlee and Lockner, 1977; Itabura and Sato, 1998); effects of thermal and stress cycling (Atkinson et al., 1984); Mori et al., 1994; Zuberek and Zogala, 1996); basic rock and coal fracture (Atkinson, 1979; Dunning and Dunn, 1980; Khair and Jung, 1989; Lockner and Byerlee, 1995); Coal cutting (Hardy and Shen, 1996; Shen, 1998); Kaiser effect studies (Hardy, 1996; Holcomb, 1993a, 1993b; Li, 1998; Momayez and Hassani, 1992; Wood et al., 1992); and a variety of other studies associated with various mechanical properties of geologic materials (e.g. Christesu, 1989; Glaser and Nelson, 1990; Itakura et al., 1995; and Rao et al., 1995). The forgoing represents a very limited outline of the recent laboratory studies underway in the AE/MS area.

REFERENCES

Albright, J.N. and Pearson, C. 1984. Microseismic Activity Observed During Decompression on an Oil Storage Cavern in Rock Salt, *Proceedings Third Conference on Acoustic Emission/Microseismic Activity in Geologic Structures and Materials*, The Pennsylvania State University, October 1981, Trans Tech Publications, Clausthal-Zellerfeld, Germany, pp. 199-210.

Anon. 1978. Annual Report – 1977, Hot Dry Rock Geothermal Energy Development Project, *Los Alamos Scientific Laboratory Progress Report LA-7109-PR*, Los Alamos, New Mexico, February 1978.

Armstrong, B.H. and Stierman, D.J. 1989. Acoustic Emission from Foreshocks and Secular Strain Changes Prior to Earthquakes, *Proceedings Fourth Conference on Acoustic Emission/Microseismic Activity in Geologic Structures and Materials*, The Pennsylvania State University, October 1985, Trans Tech publications, Clausthal-Zellerfeld, Germany, pp. 309-326.

Antsyferov, M.S. (ed.) 1966. *Seismo-Acoustic Methods in Mining*, Consultants Bureau, New York.

Asanuma, H. and Niitsuma, H. 1995. Triaxial Seismic Measurement While Drilling and Estimation of Subsurface Structure, *Geotherm. Sci. Tech., Vol. 5*, pp. 31-51.

Atkinson, B.K. 1979. A Fracture Mechanics Study of Subcritical Tensile Cracking of Quartz in Wet Environments, *Pure and Applied Geophysics, Vol. 117*, pp. 1011-1024.

Atkinson, B.K., MacDonald, D. and Meredith, P.G. 1984. Acoustic Response of Westerley Granite During Temperature and Stress Cycling Experiments, *Proceedings Third Conference on Acoustic Emission/Microseismic Activity in Geologic Structures and Materials*, The Pennsylvania State University, October 1981, Trans Tech Publications, Clausthal-Zellerfeld, Germany, pp. 5-18.

Ballivy, G., Rhazi, J.E., Bouja, A. and Li, X.J. 1998. AE During Anchor Pull-Out Tests, *Proceedings Sixth Conference on Acoustic Emission/Microseismic Activity in Geologic Structures and Materials*, The Pennsylvania State University, June 1996, Trans Tech Publications, Clausthal-Zellerfeld, Germany, pp. 4-14.

Baria, R., Hearn, K. and Batchelor, A.S. 1989. Induced Seismicity During the Hydraulic Stimulation of a Potential Hot Dry Rock Geothermal Reservoir, *Proceedings Fourth Conference on Acoustic Emission/Microseismic Activity in Geologic Structures and Materials*, The Pennsylvania State University, October 1985, Trans Tech Publications, Clausthal, Germany, pp. 327-352.

Barron, K. 1970. Detection of Fracture Initiation in Rock Specimens by the Use of a Simple Ultrasonic Listening Device, *Int. J. Rock Mechanics Min. Sci., Vol. 8*, pp. 55-59.

Bassim, M.N. 1987. Macroscopic Orgins of Acoustic Emission, *Nondestructive Testing Handbook*, Vol. 5: Acoustic Emission Testing, Section 2, P. McIntire (ed.). American Society for Nondestructive Testing, Columbus, Ohio, pp. 46-61.

Batchelor, A.S., Baria, R. and Hearn, K. 1983. *Monitoring the Effects of Hydraulic Stimulation by Microseismic Event Location: A Case Study*, Preprint No. 12109, Society of Petroleum engineers, 58th Annual Technical Conference, San Francisco, October 1983.

Bath, M. 1974. *Spectral Analysis in Geophysics*, Elsevier Scientific Publishing Company, New York.

Beard, F.D. 1962. Microseismic Forecasting of Excavation Failures, *Civil Engineering, Vol. 32*, No. 5, pp. 50-51.

Blackman, R.B. and Tukey, J.W. 1958. *The Measurement of Power Spectra*, Dover Publications, New York.

Blake, W. and Leighton, F. 1970. Recent Developments and Applications of the Microseismic Method in Deep Mines, *Proceedings Eleventh Symposium on Rock Mechanics*, (Berkeley 1969), AIME, New York, pp. 429-443.

Broadbent, C.D. and Armstrong, C.W. 1960. Design and Application of Microseismic Devices, *Proceedings Fifth Canadian Rock Mechanics Symposium*, (Toronto 1968), Department of Energy, Mines and Resources, Ottawa, pp. 91-103.

Brown, J.W. 1965. *An Investigation of Microseismic Activity in Rock Under Tension*, M.S. Thesis, Mining Department, The Pennsylvania State University.

Brown, J.W. and Singh, M.M. 1966. An Investigation of Microseismic Activity in Rock Under Tension, *Trans. Soc. Mining Eng., Vol. 233*, pp. 255-256.

Buchheim, W. (1958). *Geophysical Methods for the Study of Rock Pressure in Coal and Potash-Salt Mining*, International Strata Control Congress, (Leipzig 1958), pp. 222-223.

Byerlee, J.D. and Lockner, D. 1977. Acoustic Emission During Fluid Injection, *Proceedings First Conference on Acoustic Emission/Microseismic Activity in Geologic Structures and Materials*, The Pennsylvania State University, June 1975, Trans Tech Publications, Clausthal-Zellerfeld, Germany, pp. 87-98.

Carabelli, E., Frederici, P., Graziano F. and Sampaolo, A. 1989. "AE/MS in a Dam Area: A Study in South Italy", *Proceedings Fourth Conference on Acoustic Emission/Microseismic Activity in Geologic Structures and Materials*, The Pennsylvania State University, October 1985, Trans Tech Publications, Clausthal-Zellerfeld, Germany.

Chugh, Y.P., Hardy, H.R. Jr. and Stefanko, R. 1972. An Investigation of the Frequency Spectra of Microseismic Activity in Rock Under Tension, *Proceedings Tenth Rock Mechanics Symposium*, (Austin 1968), AIME, New York, pp. 73-113.

Christescu, N. 1989. A Mechanical-AE/MA Correlation, *Proceedings Fourth Conference on Acoustic Emission/Microseismic Activity in Geologic Structures and Materials*, The Pennsylvania State University, October 1985, Trans Tech Publications, Clausthal-Zellerfeld, Germany, pp. 559-567.

Cook, N.G.W. 1963. The Seismic Location of Rockbursts, *Proceedings 5th Rock Mechanics Symposium*, Pergamon Press, Oxford, pp. 49-80.

Crandell, F.J. 1955. Determination of Incipient Roof Failures in Rock Tunnels by Microseismic Detection, *Journal Boston Society Civil Engineers*, January 1955, pp. 39-54.

Drouillard, T.F. 1979. *Acoustic Emission: A Bibliography with Abstracts*, Plenum Publishing Company, New York, 787 pp.

Drouillard, T.F. 1996. A History of Acoustic Emission, *Journal of Acoustic Emission, Vol. 14*, No. 1, pp. 1-34.

Dunegan, H.L. 1996. Prediction of Earthquakes with AE/MS? Why Not, *Proceedings, Sixth Conference on Acoustic Emission/Microseismic Activity in Geologic Structures and Materials*, Trans Tech Publications, Clausthal-Zellerfeld, Germany, pp. 507-519.

Dunning, J. and Dunn, D. 1980. Microseismicity of Stable Crack Propagation in Quartz, *Proceedings Second Conference on Acoustic Emission/Microseismic Activity in Geologic Structures and Materials*, The Pennsylvania State University, November 1978, Trans Tech Publications, Clausthal-Zellerfeld, Germany, pp. 1-10.

Fonseka, G.M., Murrell, S.A.F. and Barnes, P. 1985. Scanning Electron Microscope and Acoustic Emission Studies of Crack Development in Rocks, *Int. J. Rock Mech. Min. Sci. & Geomech. Abst., Vol. 22*, No. 5, pp. 273-289.

Ge, M., and Hardy, H.R. Jr. 1988. The Mechanism of Array Geometry in the Control of AE/MS Source Location Accuracy, *Proceedings 29th U.S. Symposium on Rock Mechanics*, Minneapolis, Minnesota, June 1988, A.A. Balkema, Rotterdam, pp. 597-605.

Ge, M. and Mottahed, P. 1993. An Automatic Data Analysis and Source Location System, Proceedings 3rd International Symp. Rockbursts and Seismicity in Mines, Kingston, Canada, August 1993, *Rock Bursts and Seismicity in Mines*, A.A. Balkema, Rotterdam, pp. 343-348.

Glaser, S.D. and Nelson, P.R. 1990. Correlation of Quantitative AE Waveforms with Discrete Fracture Mechanisms in Rock, *Proceedings 10th International Acoustic Emission Symposium*, The Japanese Society for Non-destructive Inspection, Tokyo, pp. 414-421.

Goodman, R.E. 1963, Subaudible Noise During Compression of Rocks, *Geol. Soc. Amer. Bul., Vol. 74*, pp. 487-490.

Gowd, T.M. 1980. Factors Affecting the Acoustic Emission Response of Triaxially Compressed Rock, *Inter. J. Rock Mech. Min. Sci. & Geomech. Abst., Vol. 17*, pp. 219-223.

Haimson, B.C. and Kim, K. 1977. Acoustic Emission and Fatigue Mechanisms in Rock, *Proceedings First Conference on Acoustic Emission/Microseismic Activity in Geologic Structures and Materials*, The Pennsylvania State university, June 1975, Trans Tech Publications, Clausthal-Zellerfeld, Germany, pp. 35-56.

Hardy, H.R. Jr. 1972. Application of Acoustic Emission Techniques to Rock Mechanics Research, *Acoustic Emission*, R.G. Liptai, D.O. Harris and C.A. Tatro (eds.), STP 505, American Society for Testing and Materials, Philadelphia, Pennsylvania, pp. 41-83.

Hardy, H.R. Jr. 1980. Stability Monitoring of an Underground Gas Storage Reservoir, *Proceedings Second Conference on Acoustic Emission/Microseismic Activity in Geologic Structures and Materials*, The Pennsylvania State University, November 1978, Trans Tech Publications, Clausthal-Zellerfeld, Germany, pp. 331-358.

Hardy, H.R. Jr. 1981. Applications of Acoustic Emission Techniques to Rock and Rock Structures: A State-of-the-Art Review, *Acoustic Emissions in Geotechnical Engineering Practice*, V.P. Drnevich and R.E. Gray, (eds.), STP 750, American Society for Testing and Materials, Philadelphia, Pennsylvania, pp. 4-92.

Hardy, H.R. Jr. (ed.) 1989. *Proceedings, Fourth Conference on Acoustic Emission/ Microseismic Activity in Geologic Structures and Materials*, The Pennsylvania State University, October 1985, Trans Tech Publications, Clausthal-Zellerfeld, Germany, 711 pp.

Hardy, H.R. Jr. (ed.) 1995. *Proceedings, Fifth Conference on Acoustic Emission/ Microseismic Activity in Geologic Structures and Materials*, The Pennsylvania State University, June 1991, Trans Tech Publications, Clausthal-Zellerfeld, Germany, 755 pp.

Hardy, H.R. Jr. 1996. Application of the Kaiser Effect for the Evaluation of In-Situ Stresses in Salt, *Proceedings 3rd Conference on the Mechanical Behavior of Salt*, Ecole Polytechnique, Palaiseau, France, September 1993, Trans Tech Publications, Clausthal-Zellerfeld, Germany, pp. 85-100.

Hardy, H.R. Jr. (ed.). 1998. *Proceedings, Sixth Conference on Acoustic Emission/ Microseismic Activity in Geologic Structures and Materials*, The Pennsylvania State University, June 1996, Trans Tech Publications, Clausthal-Zellerfeld, Germany.

Hardy, H.R. Jr. and Leighton, F.W. (eds.) 1977. *Proceedings, First Conference on Acoustic Emission/Microseismic Activity in Geologic Structures and Materials*, The Pennsylvania State University, June 1975, Trans Tech Publications, Clausthal-Zellerfeld, Germany.

Hardy, H.R. Jr. and Leighton, F.W. (eds.) 1980. *Proceedings, Second Conference on Acoustic Emission/Microseismic Activity in Geologic Structures and Materials*, The Pennsylvania State University, November 1978, Trans Tech Publications, Clausthal-Zellerfeld, Germany.

Hardy, H.R. Jr. and Leighton, F.W. (eds.) 1984. *Proceedings, Third Conference on Acoustic Emission/Microseismic Activity in Geologic Structures and Materials*, The Pennsylvania State University, October 1981, Trans Tech Publications, Clausthal-Zellerfeld, Germany.

Hardy, H.R. Jr., Roberts, D.A. and Richardson, A.M. 1980. Application of Acoustic Emission in Fundamental Studies of Salt Behavior, *Proceedings Fifth International Symposium on Salt*, Vol. 1, A.H. Coogan and L. Hauber (eds.), Northern Ohio Geological Society, Inc., Cleveland, pp. 269-280.

Hardy, H.R. Jr. and Shen, H.W. 1996. Laboratory Study of Acoustic Emission and Particle Size Distributions during Linear Cutting of Coal, *Proceedings of the 2nd North American Rock Mechanics Symposium*, Montreal, June 1996, pp. 835-844.

Hardy, H.R. Jr., Kim, R.Y., Stefanko, R. and Wang, Y.J. 1970. Creep and Microseismic Activity in Geologic Materials, *Rock mechanics – Theory and Practice*, Proceedings Eleventh Symposium on Rock mechanics (Berkeley 1969), AIME, New York, pp. 377-413.

Hayashi, M., Kanagawa, T., Hibino, S., Matozina, M. and Kitahara, Y. 1979. Detection of Anisotropic Geostresses Trying by Acoustic Emission, and Non-linear Rock Mechanics on Large Excavation Caverns, *Proceedings 4th Congress International Society for Rock Mechanics*, (Montreaux 1979), A.A. Balkema, Rotterdam, Vol. 2, pp. 211-218.

Hente, B., Gommlich, G. and Flach, D. 1984. Microseismic Monitoring of Candidate Nuclear Waste Disposal Sites, *Proceedings Third Conference on Acoustic Emission/ Microseismic Activity in Geologic Structures and Materials*, Clausthal-Zellerfeld, Germany, pp. 393-401.

Holcomb, D.J. 1993a. Observations of the Kaiser Effect Under Multiaxial Stress State: Implications for its Use in Determining In Situ Stress, *Geophy. Res. Lett., Vol. 20*, No. 9, No. 2119-2122.

Holcomb, D.J. 1993b. General Theory of the Kaiser Effect, *Int. J. Rock Mech. Min. Sci. & Geomech. Abstr., Vol. 30*, No. 7, pp. 929-935.

Ishimoto, M. and Iida, K. 1939. Observations sur les Seismes Enregistres par le Microsismographe Construit Dermerment, *Bull. Earthquake Research Institute, Vol. 17*, Tokyo University, pp. 443-478.

Itakura, K. and Sato, K. 1998. AE Activity and Its Fractal Properties with Hydrofracturing of Rock Specimens, *Proceedings Sixth Conference on Acoustic Emission/Microseismic Activity in Geologic Structures and Materials*, The Pennsylvania State University, June 1996, Trans Tech Publication, Clausthal-Zellerfeld, Germany.

Itakura, K., Sato, K., Nagano, K. and Kusano, Y. 1995. Fractals on Acoustic Emission During Hydraulic Fracturing, *J. Acoustic Emission, Vol. 13*, No. 3-4, pp. S75-S82.

Khair, A.W. 1972. *Failure Criteria Applicable to Pressurized Cavities in Geologic Materials Under In-Situ Stress Conditions*, Ph.D. Thesis, Department of Mineral Engineering, The Pennsylvania State University.

Khair, A.W. 1984. Acoustic Emission Pattern: An Indicator of Mode of Failure in Geologic Materials as Affected by their Natural Imperfections, *Proceedings Third Conference on Acoustic Emission/Microseismic Activity in Geologic Structures and Materials*, The Pennsylvania State University, October 1981, Trans Tech Publications, Clausthal-Zellerfeld, Germany, pp. 45-66.

Khair, A.W. and Jung, S. 1989. Identification of Fracture in Coal by AE in Dynamic Tests, *Proceedings Fourth Conference on Acoustic Emission/Microseismic Activity in Geologic Structures and Materials*, The Pennsylvania State University, October 1985, Trans Tech Publications, Clausthal-Zellerfeld, Germany, pp. 57-72.

Koerner, R.M., McCabe, W.M. and Lord, A.E. Jr. 1981. Acoustic Emission Behavior and Monitoring of Soils, *Acoustic Emissions in Geotechnical Engineering Practice*, V. P. Drnevich and R. E. Gray, (eds.), STP 750, American Society for Testing and Materials, Philadelphia, Pennsylvania, pp. 93-141.

Koopmans, L.H. 1974. *The Spectral Analysis of Time Series*, Academic Press, New York.

Li, C. 1998. A Theory for the Kaiser Effect in Rock and Its Potential Applications, *Proceedings Sixth Conference on Acoustic Emission/Microseismic Activity in Geologic Structures and Materials*, The Pennsylvania State University, June 1996, Trans Tech Publications, Clausthal-Zellerfeld, Germany, pp. 171-185.

Lockner, D. A. and Byerlee, J.D. 1977. Acoustic Emission and Fault Location in Rocks, *Proceedings First Conference on Acoustic Emission/Microseismic Activity in Geologic Structures and Materials*, The Pennsylvania State University, June 1975, Trans Tech Publications, Clausthal-Zellerfeld, Germany, pp. 99-107.

Lockner, D.A. and Byerlee, J.D. 1995. Precursory AE Patterns Leading to Rock Fracture, *Proceedings Fifth Conference on Acoustic Emission/Microseismic Activity in Geologic Structures and Materials*, The Pennsylvania State University, June 1991, Trans Tech Publications, Clausthal-Zellerfeld, Germany, pp. 45-58.

Lord, A.E. Jr. and Koerner, R.M. 1979. Acoustic Emission in Geologic Materials, *Fundamentals of Acoustic Emission*, K. Ono (ed.), School or Engineering and Applied Science, University of California, Los Angeles, California, pp. 261-307.

Mae, I. and Nakao, K. 1968. Characteristics in the Generation of Microseismic Noises in Rocks Under Uniaxial Compressive Load, *Jour. Soc. Materials Science Japan, Vol. 17*, No. 181, pp. 62-17.

Majer, E.L., King, M.S. and McEvilly, R.V. 1984. The Application of Modern Seismological Methods to Acoustic Emission Studies in a Rock Mass Subjected to Heating, *Proceedings Third Conference on Acoustic Emission/Microseismic Activity in Geologic Structures and Materials*, The Pennsylvania State University, October 1981, Trans Tech Publications, Clausthal-Zellerfeld, Germany, pp. 499-516.

McCabe, W.M. 1980. Acoustic Emission in Coal: A Laboratory Study, *Proceedings Second Conference on Acoustic Emission/Microseismic Activity in Geologic Structures and Materials*, The Pennsylvania State University, November 1978, Trans Tech Publications, Clausthal-Zellerfeld, Germany, pp. 35-53.

McKavanagh, G.M. and Enever, J.R. 1980. Developing a Microseismic Outburst Warning System, *Proceedings Second Conference on Acoustic Emission/ Microseismic Activity in Geologic Structures and Materials*, The Pennsylvania State University, November 1978, Trans Tech Publications, Clausthal-Zellerfeld, Germany, pp. 213-225.

Meister, D. 1980. Microacoustic Studies in Rock Salt, *Proceedings Second Conference on Acoustic Emission/Microseismic Activity in Geologic Structures and Materials*, The Pennsylvania State University, November 1978, Trans Tech Publications, Clausthal-Zellerfeld, Germany, pp. 259-275.

Merrill, R.H. 1954. *Design of Underground Mine Openings, Oil Shale Mine, Rifle, Colorado*, RI 5089, USBM.

Merrill, R.H.. 1957. *Roof Span Studies in Limestone*, RI 5940, USBM.

Merrill, R.H. and Morgan, T.A. 1958. *Method of Determining the Strength of a Mine Roof*, RI 5406, USBM.

Milne, W.G. (ed.) 1976. *Induced Seismicity*, Special Issue, *Engineering Geology, Vol. 10*, No. 2-4, pp. 83-388.

Mogi, K. 1962. Study of the Elastic Shocks Caused by the Fracture of Heterogeneous Materials and its Relation to Earthquake Phenomena, *Bulletin Earthquake Research Institute, Vol. 40*, pp. 125-173.

Mogi, K. 1968. Source Location of Elastic Shocks in the Fracturing Process in Rocks, *Bulletin of the Earthquake Research Institute, Vol. 46*, pp. 1103-1125.

Momayez, M. and Hassani, F.P. 1992. Application of Kaiser Effect to Measure In-situ Stress in Underground Mines, *Proceedings 33rd US Symposium on Rock Mechanics*, Santa Fe, June 1992, A.A. Balkema, Rotterdam, pp. 979-988.

Montoto, M., Suárez del Río, L.M., Khair, A.W. and Hardy, H.R. Jr. 1984. AE in Uniaxially Loaded Granitic Rocks in Relation to Their Petrographic Character, *Proceedings Third Conference on Acoustic Emission/Microseismic Activity in Geologic Structures and Materials*, The Pennsylvania State University, October, 1981, Trans Tech Publications, Clausthal-Zellerfeld, Germany, pp. 83-100.

Nagano, K., Saito, H. and Niitsuma, H. 1995. Guided Waves Trapped in an Artificial Subsurface Fracture, *Geothermal Sci. Technol., Vol. 5*, pp. 63-70.

Nagano, K., Sato, K. and Niitsuma, H. 1998. Fracture Orientation Estimated from 3C Crack Waves, *Proceedings Sixth Conference on Acoustic Emission/Microseismic Activity in Geologic Structures and Materials*, The Pennsylvania State University, June 1996, Trans Tech Publications, Clausthal-Zellerfeld, Germany.

Niitsuma, H., Nakatsuka, K., Chubachi, N., Yokoyama, H. and Takanohashi, M. 1984. New Evaluation Method of Geothermal Reservoir by Field Measurement,

Proceedings 7th International Acoustic Emission Symposium, Zao, Japan, October 1984, Japanese Society for Non-Destructive Inspection, Tokyo, pp. 642-651.

Mori, Y., Saruhaski, K. and Mogi, K. 1994. Acoustic Emission from Rock Specimen Using Cyclic Loading, *Progress in Acoustic Emission VII*, Proceedings 12th International Acoustic Emission Symposium, Sapporo, Japan, October 1994, Japanese Society for Non-Destructive Inspection, Tokyo, pp. 173-178.

Obert, L. 1977. The Microseismic Method: Discovery and Early History, *Proceedings First Conference on Acoustic Emission/Microseismic Activity in Geologic Structures and Materials*, The Pennsylvania State University, June 1975, Trans Tech Publications, Clausthal-Zellerfeld, Germany, pp. 11-12.

Obert, L. and Duvall, W.I. 1942. *Use of Subaudible Noises for the Prediction of Rock Bursts, Part II*, RI 3654, USBM.

Obert, L. and Duvall, W.I. 1945. *Microseismic Method of Predicting Rock Failure in Underground Mining, Part II, Laboratory Experiments"*, RI 3803, USBM.

Ohtsu, M. 1994. Application of Civil Engineering in Japan with the Emphasis on Concrete and Soil Mechanics, *Progress in Acoustic Emission VII*, Proceedings 12th International Acoustic Emission Symposium, Sapporo, Japan, October 1994, The Japanese Society for Non-Destructive Inspection, Tokyo, pp. 51-58.

Paulsen, J.C., Kistler, R.B. and Thomas, L.L. 1967. Slope Stability Monitoring at Boron, *Mining Congress Journal, Vol. 53*, pp. 28-32.

Pérami, R. and Thénoz, B. 1969. Comparison des Compartments de Divers Granites Soumis à des essays Uniaxiaux de Microfissuration, *Revue De L'Industrie Minerale*, July 1969, pp. 50-62.

Rao, M.V.M.S., Nishizawa, O. and Kusunose, K. 1995. Microcrack Damage in Brittle Rock: A Case Study, *Proceedings Fifth Conference on Acoustic Emission/Microseismic Activity in Geologic Structures and Materials*, The Pennsylvania State University, June 1991, Trans Tech Publications, Clausthal-Zellerfeld, Germany, pp. 73-80.

Reymond, M.C. 1975. Mechanism of Brittle Fracture of Rock During Uniaxial Compression in Laboratory and in a Quarry by an Acoustic Emission Technique, (In French), *Rock Mechanics, Vol. 7*, pp. 1-16.

Robinson, E.A. and Dussani, T.S. 1986. *Geophysical Signal Processing*, Prentice Hall International, Englewood Cliffs, N.J., 481 pp.

Savage, J.C. and B.B. Mohanty (1969). Does Creep Cause Fracture in Brittle Rocks, *J. Geophys. Res., Vol. 74*, No. 17, pp. 4329-4332.

Scholz, C.H. 1968a. The Frequency-Magnitude Relation of Microfracturing in Rock and Its Relation to Earthquakes, *Bull. Seismol. Soc. Am.*, Vol. 58, pp. 399-415.

Scholz, C.H. 1968b. "Experimental Study of the Fracturing Process in Brittle Rock", *Jour. Geoph. Res., Vol. 73*, p. 1447.

Scholz, C.H. 1968c. Microfracturing and Inelastic Deformation of Rock in Compression, *Jour. Geoph. Res., Vol. 73*, p. 1417.

Shen, H.W. 1998. Application of Acoustic Emission Techniques to the Monitoring of Cutting and Breakage of Geologic Materials, *Proceedings Sixth Conference on Acoustic Emission/Microseismic Activity in Geologic Structures and Materials*, The Pennsylvania State University, June 1996, Trans Tech Publications, Clausthal-Zellerfeld, Germany, pp. 85-107.

Sondergeld, C.H., Estey, L.H. and Granryd, L. 1984. Acoustic Emissions During Compression Testing of Rock, *Proceedings Third Conference on Acoustic Emission/Microseismic Activity in Geologic Structures and Materials*, The Pennsylvania State University, October 1981, Trans Tech Publications, Clausthal-Zellerfeld, Germany, pp. 131-145.

Stead, D. and Szczepanik, Z. 1995. Acoustic Emission During Uniaxial Creep in Potash, *Proceedings Fifth Conference on Acoustic Emission/Microseismic Activity in Geologic Structures and Materials*, The Pennsylvania State University, June 1991, Trans Tech Publications, Clausthal-Zellerfeld, Germany, pp. 97-111.

Suzuki, Z. 1953. A Statistical Study on the Occurrence of Small Earthquakes – I, Science Reports Tohoku University, *Fifth Series Geophysics, Vol. 5*, pp. 177-182.

Suzuki, Z. 1954. "A Statistical Study on the Occurrence of Small Earthquakes – II", Science Reports Tohoku University, *Fifth Series Geophysics, Vol. 6*, pp. 105-118.

Suzuki, T., Sasaki, T. and Hirota, T. 1964. A New Approach to the Prediction of Failure by Rock Noise, *Fourth International Conference on Strata Control and Rock Mechanics*, Columbia University, New York, pp. 1-9.

Unal, E. and Hardy, H.R. Jr. 1984. Application of AE Techniques in the Evaluation of Rock Bolt Anchor Stability, *Proceedings Third Conference on Acoustic Emission/Microseismic Activity in Geologic Structures and Materials*, The Pennsylvania State University, October 1981, Trans Tech Publications, Clausthal-Zellerfeld, Germany, pp. 173-196.

Wadley, H.N.G. 1987. Microscopic Origins of Acoustic Emission, *Nondestructive Testing Handbook*, Vol. 5: Acoustic Emission Testing, Section 3, (Editor, P. McIntire), American Society for Nondestructive Testing, Columbus, Ohio, pp. 64-90.

Wisecarver, D.W., Merrill, R.H. and Stateham, R.M. 1969. The Microseismic Technique Applied to Slope Stability, *Transactions Society of Mining Engineers, Vol. 244*, pp. 378-385.

Wood, B.R.A., Harris, R.W. and Colburn, R.J. 1984. Long Term Monitoring of a Large Concrete Dam with Acoustic Emission, *Proceedings 7th International Acoustic Emission Symposium*, Zao, Japan, October 1984, Japanese Society for Non-Destructive Inspection, Tokyo, pp. 642-651.

Wood, B.R.A., Harris, R.W., Porter, J.B. and Mete, T. 1992. An Evaluation of the Kaiser Effect and Wave Propagation Characteristics in Rock Samples, *Progress in Acoustic Emission VI*, Proceedings 11th International Acoustic Emission Symposium, Fukuoka, Japan, October 1992, The Japanese Society for Non-Destructive Inspection, Tokyo, pp. 183-193

Young, R.P., Hutchins, D.A., McGaughey, W.J., Urbanic, T., Falls, S. and Towers, J. 1987. Concurrent Seismic Tomographic Imaging and Acoustic Emission Techniques: A New Approach to Rockburst Investigations, *Proceedings Sixth International Congress on Rock Mechanics*, Montreal, Canada, August/September 1987, Vol. 2, pp. 1333-1338.

Zuberek, W.M. and Zogala, B. 1996. Memory of Maximum Stresses and Maximum Temperatures During Rock Deformation, *Inst. Geoph. Pol. Acad. Sci., M-18*, (273), pp. 133-138.

CHAPTER 2

Energy transmission and detection

2.1 INTRODUCTION

Energy from an AE/MS source (seismic source) propagates outward from the
source in the form of an elastic stress wave. These waves may be detected by
surface or subsurface transducers which sense the particle motion associated
with the elastic wave. Figure 2.1 illustrates a highly simplified field case
where the medium is assumed to be isotropic, homogeneous and continuous.
Depending on the source mechanism and the boundary conditions, energy may
be transmitted in the form of plane, cylindrical or spherical waves.

 A similar situation occurs in the laboratory where specimens are instrumented
in order to monitor the AE/MS activity associated with a particular test program.
In such cases surface mounted AE/MS transducers are normally employed and
these sense the particle motion associated with the area of the specimen surface
to which they are attached.

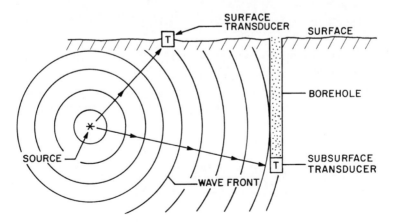

Figure 2.1. Propagation of AE/MS source energy to a surface or subsurface
transducer.

In both field and laboratory studies the particle motion at the transducer is dependent on a number of factors including the form and energy level of the source, the geometry of the test structure, the mechanical properties of the associated media, and the type and degree of discontinuities present. The preceding factors all have a direct or indirect influence on the propagation of the elastic waves generated by the source and on a variety of factors which influence the generation and propagation of these waves, namely: the type of wave propagation, propagation velocities, wave attenuation, and reflection and refraction at boundaries and interfaces. A detailed analysis of these various factors is considered outside the scope of the present text; however, a number of the more important factors will be briefly discussed in order to provide the reader a better appreciation of the processes which occur between the structural instability under study (source) and the actual monitoring site (transducer location).

2.2   WAVE PROPAGATION

As noted earlier, propagation of energy from a seismic source may occur by means of a variety of types of waves, namely plane, cylindrical and spherical waves. In the following section some important characteristics of these waves will be briefly considered. Further details on wave propagation in general and the associated mathematical analyses are available in a number of other texts, including Dobin (1960), Dowding (1985), Dresen et al. (1994), Ewing et al. (1957), Gibowicz and Kijko (1994), Kolsky (1963), Paillet and Cheng (1991), Pierce (1981) and Reinhart, (1975). Further information relative to wave propagation, specifically associated with AE/MS phenomena, is presented by Egle (1987).

2.2.1   Plane Waves

True plane waves can only be generated and propagated in an unbounded media, for example, by planar impact of a long rod as illustrated in Figure 2.2.

The particle motion associated with a plane wave is either perpendicular or parallel to the wave front, depending on whether it is a P-wave or an S-wave.

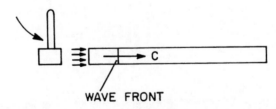

WAVE FRONT

Figure 2.2. Generation of a plane wave in a long rod.

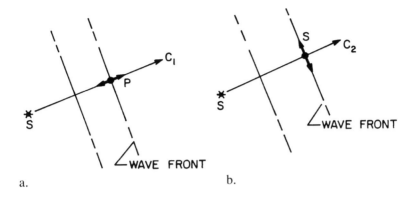

Figure 2.3. Particle motion associated with plane waves. a. P-wave. b. S-wave.

The wave fronts associated with these two types of waves move at two different velocities, C_1 and C_2, respectively, where $C_1 > C_2$.

In an extended media, however, seismic sources generate either cylindrical or spherical waves. These waves change their shape as they advance, in contrast to plane waves. In the "far field", i.e., a relatively long distance from the source, the wave front of both cylindrical and spherical waves may be assumed, over small distances along the wave front, to be planer in form.

2.2.2 Cylindrical waves

Figure 2.4 illustrates the general form of a cylindrical wave source (s). An example would be a long column of explosive located in a borehole and detonated instantaneously along its length. Here the resulting wave front associated with the P- and S-wave components advance outwards through the surrounding material as an expanding cylindrical shell. Figure 2.4b illustrates the case for the P-wave component. In the far field situation a short segment of the wave front may be treated as if it were a plane wave.

As the wave spreads out from its source the associated energy must be distributed over an area that increases as the radius of the cylindrical wave front (surface area = $2\pi rl$). Since the energy in the elastic wave is proportional to the square of its amplitude (i.e., $\varepsilon \propto A^2$) and the energy per unit area is decreasing as the surface area increases, then due to geometric spreading only, the wave amplitude will fall off as $1/r^{1/2}$ namely:

$$A = A_o / r^{1/2} \qquad\qquad (\text{Eq. } 2.1)$$

where A is the wave amplitude at a radius r from the source, A_o is the wave amplitude at the source and r is the distance from the source.

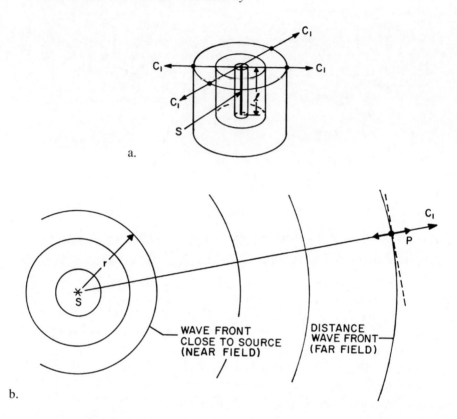

Figure 2.4. Cylindrical source and associated wave propagation. a. Cylindrical source. b. Wave propagation from a cylindrical source showing the P-wave component.

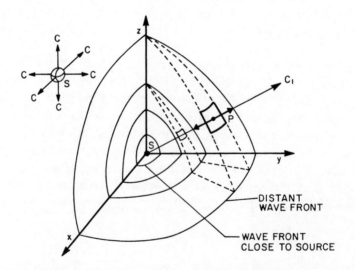

Figure 2.5. General form of a spherical source and associated P-wave propagation.

2.2.3 Spherical waves

Figure 2.5 illustrates the general form of a spherical wave source. An example of such a source could be a small spherical explosive charge. Here the resulting P- and S-wave fronts advance outwards through the surrounding material as an expanding spherical shell. In the x-y plane the wave propagation is similar to that shown earlier in Figure 2.4b.

As in the case of the cylindrical wave the energy per unit area decreases with distance from the source. In the case of the spherical wave the surface area of the wave front increases as the square of the radius (surface area $= 4\pi r^2$) and, due to geometric spreading only, the observed signal amplitude will fall off as 1/r, namely:

$$A = A_o / r \qquad \text{(Eq. 2.2)}$$

2.3   BODY AND SURFACE WAVES

There are two basic types of elastic waves, namely: body waves, which travel through the interior of a material, and surface waves which travel along the surface of a material or, under special conditions, along a bounded layer.

2.3.1   Body waves

Body waves are progressive excitations of ground volume elements involving elastic dilations and distortions as shown in Figure 2.6. In compressional waves (Fig. 2.6a) the particle motions are in the direction of propagation. Compressional waves are also called longitudinal waves, dilatational waves, pressure waves, irrotational waves, primary or P-waves. The term P-wave will be utilized generally throughout this text.

In shear waves (Fig. 2.6b) the particle motions are perpendicular to the direction of propagation. Shear waves are also called transverse waves, distortional waves, secondary or S-waves. The term S-wave will be utilized generally throughout this text. When shear wave energy is polarized so that both particle motion and wave propagation are in the vertical plane, the shear wave components are denoted by SV. On the other hand, if particle motion is polarized horizontally and transversely to wave propagation in the vertical plane, the components are denoted by SH. P- and S-waves propagate at different velocities, namely C_1 and C_2, respectively. These velocities depend on the properties of the medium, however, in general $C_2 \approx 0.6\, C_1$.

2.3.2   Surface waves

In contrast to body waves, surface waves (interface waves, guided waves) are in principle plane waves where the strain energy travels along the material

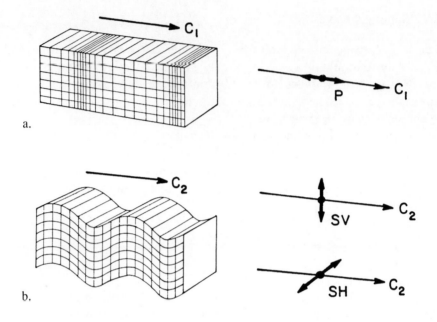

Figure 2.6. Particle motions associated with P- and S-type body waves (after Borm, 1978). a. P-waves. b. S-waves.

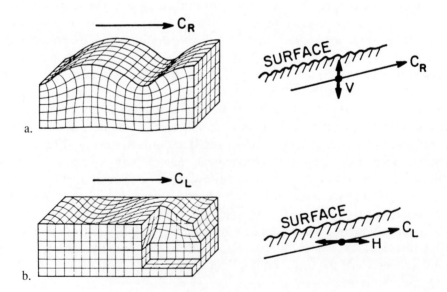

Figure 2.7. Particle motions associated with rayleigh and love type surface waves (after Borm, 1978). a. Rayleigh waves. b. Love waves

surface. Surface waves include Rayleigh waves and Love waves. In Rayleigh waves (Fig. 2.7a) the particle motion is elliptical and retrograde with respect to propagation and vertically polarized (surface SV-waves). In Love waves (Fig. 2.7b) particle motion is horizontally polarized and transverse to propagation (surface SH-waves).

Since both Rayleigh and Love waves spread out in only two directions (horizontal plane) their amplitude falls-off more slowly than body waves and as a result they can often be observed over large distances. As shown in Figure 2.8, they often have large amplitudes with respect to the P- and S-waves associated with a specific event. Typical surface wave velocities are approximately 0.92 C_2 and such waves therefore arrive after the S-waves.

A special type of Love wave was investigated by Stoneley which is propagated along an internal stratum bounded on both sides by thick layers of material which differ from it in elastic properties. Such waves, termed Stonely or channel waves, are observed in coal seams where the elastic properties of the coal differ greatly from those of the roof and floor rock, and are characterized by very low attenuation. Figure 2.9 illustrates the direction of particle motion associated by Stoneley waves.

2.4 SEISMIC SOURCES

Field and laboratory techniques for the generation of elastic waves (seismic sources) for site or specimen characterization and for AE.MS monitoring system calibration will be discussed later in this book. However, in order to

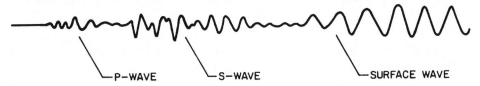

Figure 2.8. Typical time series associated with a specific event showing the associated P-, S- and surface waves.

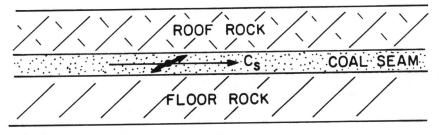

Figure 2.9. Particle motion associated with stoneley waves.

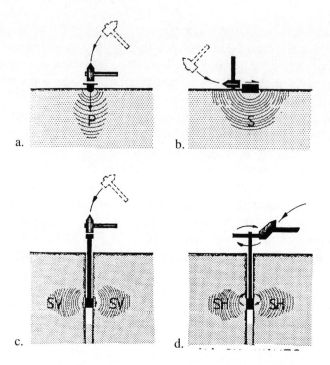

Figure 2.10. Simplified field methods for generating various types of seismic waves (after Borm, 1978). a. P-waves. b. S-waves. c. SV-waves. d. SH-waves.

provide the reader a better appreciation of various source mechanisms and the associated wave propagation, Figure 2.10 illustrates simplified field methods for generating various types of waves including P-, S-, SV- and SH-waves. More sophisticated seismic sources including explosives, "sparkers", air guns and electro-hydraulic, piezo-electric and magneto-strictive devices are also utilized. A number of these are capable of repetitive stress wave generation.

A variety of small-scale seismic sources have also been developed for use in the laboratory. These include methods based on pencil lead and glass capillary fracture, the impact of small ball bearings, the helium jet, and devices incorporating piezoelectric elements. The paper by Breckenridge et al. (1990) provides an excellent description of these methods.

2.5   SEISMIC VELOCITIES

2.5.1   Calculation of seismic velocities

The velocity of an elastic wave depends mainly on compressibility and shear characteristics of the propagation medium (e.g., a geological formation). Based on the assumption of isotropic and homogeneous elasticity of the medium, the velocity of compressional elastic body waves (C_1) is given by

$$C_1 = \sqrt{(\lambda + 2\mu)/\rho} \qquad \text{(Eq. 2.3)}$$

and the velocity of the shear waves (C_2) is given by

$$C_2 = \sqrt{\mu/\rho} \qquad \text{(Eq. 2.4)}$$

where λ and μ are the first and second Lamé constants, and ρ is the mass density.

Considering surface waves, in most cases of practical interest the Rayleigh wave velocities do not differ significantly from body shear wave velocities, whereas the velocities of Love waves vary between those of body compressional and shear waves depending on the shallow ground conditions.

In many engineering applications the elastic moduli, Young's modules (E) and Poisson's ratio (υ), of the associated media are available. These are related to the Lamé constants by the following relationships:

$$E = \mu(3\lambda + 2\mu)/(\lambda + \mu) \qquad \text{(Eq. 2.5)}$$

$$\upsilon = 0.5\,\lambda/(\lambda + \mu) \qquad \text{(Eq. 2.6)}$$

Therefore Equations 2.3 and 2.4 may be rewritten in terms of E and υ as follows:

$$C_1 = \sqrt{\frac{E}{\rho}\left\{\frac{1-\upsilon}{(1-2\upsilon)(1+\upsilon)}\right\}} \qquad \text{(Eq. 2.7)}$$

$$C_2 = \sqrt{\frac{E}{\rho}\left\{\frac{1}{2(1+\upsilon)}\right\}} \qquad \text{(Eq. 2.8)}$$

and the ratio of the P- and S-wave velocities is given by the following

$$C_1/C_2 = \sqrt{\frac{2(1-\upsilon)}{(1-2\upsilon)}} \qquad \text{(Eq. 2.9)}$$

Using Equation 2.9, for example, for $\upsilon = 0.3$, $C_1/C_2 = 1.97$ or $C_2/C_1 = 0.53$ indicating that the S–wave velocity is approximately 1/2 of the P-wave velocity.

When the medium is anisotropic, for example, a bedded deposit where the elastic moduli parallel and perpendicular to the bedding plane are significantly different, seismic velocities will depend on the direction of propagation.

Furthermore it is important to note that even in the case of isotropic media, the seismic velocities computed on the basis of laboratory values of the elastic moduli, or determined by laboratory scale tests, may not provide meaningful velocity data for field use.

2.5.2   Typical seismic velocities

In general, the two basic wave velocities, the P-wave velocity, C_1, and the S-wave velocity, C_2, are of the order of a few thousand meters per second. Such waves will move through a distance of one meter in about 200 microseconds. Typical values of the velocities C_1 and C_2 presented in Table 2.1 are only representative ones since they are sensitive to the state of stress, temperature, composition, mechanical history, and mechanical state of each material. Generally in metals, wave velocities are relatively insensitive to these factors, and high temperatures or high pressures are necessary to produce significant changes. In rocks, however, many of these same factors strongly influence wave velocities by consolidating the rock, closing microfractures and voids, and increasing the elasticity, thus enabling the rock to transmit energy faster.

As indicated in Table 2.1 the metals, with the exception of lead, have relatively high velocities. High velocity values are also found for a number of rocks such as granite, basalt and halite. In contrast some of the rocks, particularly coal and shale exhibit relatively low values. The widely differing velocities between materials often effects the propagation of elastic waves through a structure or specimen containing regions (e.g., layers) of material with different velocity characteristics. Such effects, which occur at the boundary between different materials, include reflection, refraction and mode conversion. A brief discussion of these effects is presented later in this chapter.

In a specific field or laboratory study it is necessary to evaluate the velocity characteristics of the specific materials involved. Where possible the materials should be evaluated under typical conditions of stress, temperature, moisture, etc. Furthermore velocity data necessary for field application should, if possible, be based on suitable field tests, rather than laboratory tests carried out on specimens obtained from the specific field site. Techniques for laboratory and field evaluation of wave velocities are discussed later in this book.

2.6   RELATIONSHIP BETWEEN WAVE AND PARTICLE VELOCITIES

2.6.1   General

Two points of view are commonly employed in viewing transient wave motion. In the first, the wave is viewed as a spatial variation of stress or particle velocity; and in the second, the stress at, or the motion of, a particular point is observed

Table 2.1. Typical values of P-wave and S-wave velocities (after Rinehart (1975) and other sources).

Material	Velocities – m/sec C_1	C_2
Aluminum	6,100	3,100
Brass	4,300	2,000
Glass	6,800	3,300
Steel	5,800	3,100
Lead	2,200	700
Plexiglass	2,600	1,300
Polystyrene	2,300	1,200
Magnesium	6,400	3,100
Water	1,485	– [1]
Air	331	– [1]
Ice	3,200	1,920 [2]
Sandstone	2,000	1,200 [2]
Limestone	3,200	1,920 [2]
Granite	5,000	3,000 [2]
Basalt	5,400	3,240 [2]
Halite	4,500	2,700 [2]
Shale	2,250	1,350 [2]
Coal	1,100	660 [2]

[1] Fluids do note support shear stress.
[2] Computed from C_1 data assuming $C_2/C_1 = 0.6$.

as a function of time. Figure 2.11 illustrates a simple triangular stress wave with a maximum amplitude of σ_0. In terms of its spatial variation, assuming that time is fixed, it would appear as in Figure 2.11a. Here λ is defined as the wavelength of the stress wave. In contrast, its temporal variation, as noted by a fixed observer, would appear as in Figure 2.11b. Here τ is defined as the period of the stress wave.

Now, as shown in Figure 2.12, consider a long bar impacted at point p. This generates a stress wave, with a form similar to that illustrated in Figure 2.11, which moves to the right along the bar with velocity C. At times t_1, t_2 and t_3 the front of the stress wave will have reached the position x_1, x_2, and x_3, respectively, and the spatial distributions will be as shown in Figure 2.11a.

Now the wave velocity (C) is defined as the speed at which the disturbed region, defined here as the front edge of the stress pulse, moves through the material; therefore,

$$C = \frac{\Delta x}{\Delta t} = \frac{x_2 - x_1}{t_2 - t_1} \qquad \text{(Eq. 2.10)}$$

As noted earlier (Equations 2.7 and 2.8), wave velocities are constants which depend on the mechanical properties of the materials involved.

Now, in Figure 2.12 consider what occurs as the stress wave passes the plane CD and reaches a position Δx further to the right. The situation is illustrated in detail in Figure 2.13. At time $t = t_1$ the stress wave has just reached the boundary CD and a short period of time (Δt) later it reaches the boundary EF a distance Δx to the right. During a period of time Δt a mass of material, $m = A(\Delta x)\rho$, has been subjected to an impulsive load. Now from Newton's second law of motion, impulse is equal to the resulting change of momentum, namely:

$$F \Delta t = \Delta M = m(V_f - V_i)$$

In the present case

$$(\sigma_x A) \Delta t = (\rho \Delta x A) (V_x - 0)$$

which reduces to

$$\sigma_x = \left[\rho \frac{\Delta x}{\Delta t}\right] V_x \qquad \text{(Eq. 2.11)}$$

From Equation 2.10, $C = \Delta x/\Delta t$, therefore, Equation 2.11 further reduces to

$$\sigma_x = \rho C V_x \qquad \text{(Eq. 2.12)}$$

or in general

$$\sigma = \rho C V \qquad \text{(Eq. 2.13)}$$

where σ is the stress, ρ is the mass density, C is the wave velocity and V is the particle velocity. Since ρ and C are material properties Equation 2.13 provides the relationship between the stress (σ) and the resulting particle velocity (V).

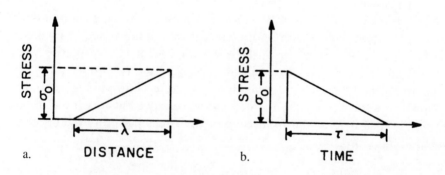

Figure 2.11. Triangular stress wave viewed in terms of its spatial and temporal variation. a. Spatial variation (time fixed, variation with distance noted). b. Temporal variation (position fixed, variation with time noted).

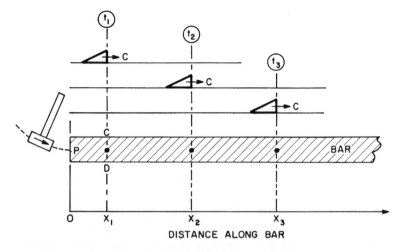

Figure 2.12. Stress wave propagation in a long bar impacted at point p. (Spatial distribution of stress wave after a time t_1, t_2 and t_3 is shown.)

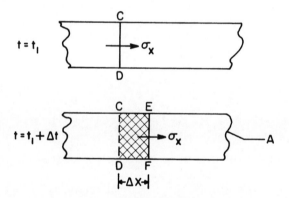

Figure 2.13. Variation of stress wave position during a time Δt.

2.6.2 Acoustic impedance

As noted in Equation 2.13 the relationship between stress, wave velocity and particle velocity is as follows:

$$\sigma = \rho CV$$

where the produce ρC is defined as the acoustic impedance (Z), namely:

$$Z = \rho C \qquad \text{(Eq. 2.14)}$$

Since the acoustic impedance depends on the mass density of the material and the value of the associated P- or S-wave velocity, an isotropic material therefore has two specific acoustic impedance values, namely:

$$Z_1 = \rho C_1 \qquad \text{(Eq. 2.15)}$$

and

$$Z_2 = \rho C_2 \qquad \text{(Eq. 2.16)}$$

Typical values of Z_1 for a variety of materials are listed in Table 2.2. As noted in Table 2.2 the larger values of Z_1 are associated with the metals, with values for rock in most cases being considerably smaller. A large variation in Z_1 values is evident within the rocks listed, ranging from 13.2 for coal to 154.4 for basalt. Since Z_1 depends directly on C_1, it is important that suitable values of C_1 be utilized to calculate the appropriate Z_1 values for a specific laboratory or field situation.

Variations in acoustic impedance within a composite structure (e.g., a layered media) have a critical influence on the transmission of AE/MS energy through such a structure due to the reflection and refraction that occurs at boundaries between areas having different values of acoustic impedance.

Table 2.2. Typical values of acoustic impedance associated with P-wave propagation.

Material	C_1 m/sec	ρ gm/cm^3	Z_1 10^4 gm/sec-cm^2
Aluminum	6100	2.7	164.7
Brass	4300	8.4	361.2
Plexiglass	2600	1.2	31.2
Steel	5800	7.8	452.4
Lead	2200	11.3	248.6
Water	1483	1.0	14.9
Air	331	0.0013	0.0043
Ice	3200	0.92	24.4
Sandstone	2000	2.1	42.0
Limestone	3200	2.3	73.6
Granite	5000	2.7	135.0
Basalt	5400	2.86	154.4
Halite	4500	2.0	90.0
Shale	2250	2.51	56.5
Coal	1100	1.2	13.2

2.7 EFFECTS AT BOUNDARIES

2.7.1 General

When an incident elastic wave strikes a cohesive boundary or interface its characteristics are modified in response to the differences in the acoustic impedance of the materials on either side of that boundary. In general the incident wave will be partly reflected from the boundary and partly transmitted through the boundary into the adjacent material. If the incident wave strikes the boundary at an oblique angle an additional phenomenon, known as "mode conversion", may also occur. Here, for example, an incident P-wave will produce both reflected and refracted P- and S-wave components. The degree of mode conversion, i.e., conversion of P-wave energy to S-wave energy, depends on the angle of incidence and the values of acoustic impedance of the materials on both sides of the boundary. Mode conversion also occurs when an S-wave is incident on a boundary at an oblique angle. It is important to note, however, that due to conservation of energy, the total energy associated with the various reflected and refracted wave components must be equivalent to the energy of the incident wave. The partition of energy and the associated amplitudes of the various wave components are dependent on the boundary conditions.

The following sections will include a brief discussion of the behavior of plane waves at cohesive and non-cohesive boundaries. This will serve to illustrate, in an approximate manner, the type of behavior expected to occur as an AE/MS wave traverses a path from the source to the monitoring transducer. It should be pointed out, however, that the reflection and refraction of non-planar waves at plane boundaries or of plane waves at non-planar boundaries usually can be treated in only a semi-quantitative way.

According to Rinehart (1975) an effective stratagem is to think of each surface or wave front as made up of an infinite number of infinitesimally small planar elements, each element tangent to the surface at a point. The problem then resolves itself into describing a set of interactions with each interaction being between only two planar elements, usually non-parallel. Since these interactions are between planar elements, equations based on planar waves can be used to calculate the partitioning of energy among the new P- and S-waves generated at each point. However, since for most curved wave fronts and curved surfaces the orientation between planar elements will vary from point to point, the relative amounts of energy going into each type of wave will also change from point-to-point.

2.7.2 Normal incidence-cohesive boundary

Although normal incidence is really a special case of oblique incidence, where the angle of incidence $\alpha = 0$, it is instructive to consider this simple situation first. Figure 2.14 illustrates the normal incidence of a compressional wave on the boundary KL between two materials with different values of acoustic impedance, namely: Z_1, Z_2, and Z_1', Z_2', where the subscripts (1 and 2) denote acoustic impedance values associated with P- and S-wave velocities. In this example it is assumed that the incident stress wave (σ_I) is of the triangular form shown earlier in Figure 2.11, with a maximum amplitude of σ_0.

For normal incidence it has been shown (Rinehart, 1975) that,

$$\sigma_I + \sigma_R = \sigma_T \qquad \text{(Eq. 2.17)}$$

and that the amplitudes of the transmitted and reflected components are as follows:

$$\sigma_T = \left(\frac{2Z_1'}{Z_1' + Z_1} \right) \sigma_1 \qquad \text{(Eq. 2.18)}$$

$$\sigma_R = \left(\frac{Z_1' - Z_1}{Z_1' + Z_1} \right) \sigma_1 \qquad \text{(Eq. 2.19)}$$

where σ_I, σ_T and σ_R are the amplitudes of the incident, transmitted and reflected waves, and Z_1 and Z_1' are the acoustic impedances of the two materials associated with compressional wave propagation.

Considering now a number of special cases relative to the values of Z_1 and Z_1'. For convenience Equations 2.18 and 2.19 may be rewritten as follows:

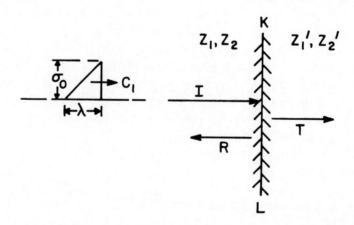

Figure 2.14. Normal incidence of a compressional wave on a boundary between two materials having different acoustic impedances.

$$\sigma_T / \sigma_1 = \left(\frac{2Z_1'}{Z_1' + Z_1} \right) \qquad \text{(Eq. 2.20)}$$

$$\sigma_R / \sigma_1 = \left(\frac{Z_1' - Z_1}{Z_1' + Z_1} \right) \qquad \text{(Eq. 2.21)}$$

Case 1 ($Z_1 = Z_1'$)

$$\sigma_T / \sigma_I = 1$$

$$\sigma_R / \sigma_I = 0$$

No reflection occurs at the boundary and $\sigma_T = \sigma_I$.

Case 2 ($Z_1 < Z_1'$)

$$\sigma_T / \sigma_I > 0$$

$$\sigma_R / \sigma_I > 0$$

Here, both reflection and transmission occur, and the magnitude of both stress waves are of the same sign as σ_I, namely compressive.

Case 3 ($Z_1 > Z_1'$)

$$\sigma_T / \sigma_I > 0$$

$$\sigma_R / \sigma_I < 0$$

Here both reflection and transmission occur. The magnitude of the transmitted wave is the same sign as σ_I, namely compressive; however, the magnitude of the reflected wave is negative in sign, indicating that it is tensile.

Case 4 ($Z_1 > 0$, $Z_1' = 0$)

$$\sigma_T / \sigma_I = 0$$

$$\sigma_R / \sigma_I = -1$$

This is the so-called "free surface" case where no transmission occurs. Here the total energy of the incident wave is reflected off the surface back into the solid as a tensile wave ($\sigma_R = \sigma_I$).

Figure 2.15 illustrates the overall variation of σ_T / σ_I and σ_R / σ_I versus Z_1' / Z_1 over the range of $0 \leq Z_1' / Z_1 \leq 10$. For $Z_1' / Z_1 > 1$, i.e. stress wave going from a low Z medium to a high Z one, both σ_T / σ_I and σ_R / σ_I are positive in sign, and as $Z_1' / Z_1 \to \infty$ the values of σ_T / σ_I and σ_R / σ_I become asymptotic to 2.0 and 1.0, respectively. For $Z_1' / Z_1 < 1$, i.e., stress wave going from a high Z medium to a low medium, $\sigma_T / \sigma_I \to 0$ and σ_R / σ_I becomes negative (tensile reflected stress) reaching a lower limit of -1 as $Z_1' / Z_1 \to 0$.

It is clear that even for the relatively simple case of normal incidence, stress wave propagation at a boundary between two media is highly sensitive to the acoustic impedances of the associated media. Since stress waves associated with AE/MS activity, in both field and laboratory situations, must often traverse a number of such boundaries as they propagate from the source to the monitoring transducer, the acoustic impedances of the associated material must be carefully considered. For example, a stress wave passing from coal

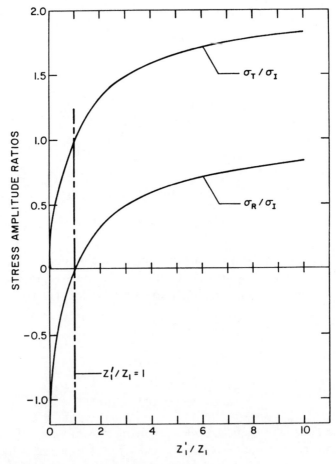

Figure 2.15. Variation of σ_T / σ_I and σ_R / σ_I with Z_1' / Z_1.

(low Z) to sandstone (high Z) will be influenced in a different manner than one passing from sandstone (high Z) to coal (low Z). The situation becomes further complicated if the stress waves strike the boundary obliquely.

2.7.3 Oblique incidence-cohesive boundary

When a longitudinal wave is incident at an oblique angle to a cohesive boundary between two dissimilar media, part of the energy will be transmitted (refracted) and part will be reflected. In contrast to the case of normal incidence, however, both longitudinal and shear wave components are involved. For example, Figure 2.16 illustrates a P-wave (P_I) incident on the boundary KL. In order to satisfy various boundary conditions at the point of contact, four stress wave components are generated, namely: a transmitted and reflected P-wave (P_T and P_R) and a transmitted and reflected S-wave (S_T and S_R).

The angular relationship between the various stress wave components are as follows:

$$\text{Sin } \alpha/C_1 = \text{Sin } \beta/C_2 = \text{Sin } \eta \; C_1' = \text{Sin } \xi/C_2' \qquad \text{(Eq. 2.22)}$$

where α, β, η and ξ are the angles associated with the propagation directions of the various stress wave components, and C_1 & C_2 and C_1' & C_2' are the P- and S-wave velocities of the materials to the left and right side of the boundary, respectively.

The fact that a P-wave striking a boundary at an oblique angle may generate not only reflected and refracted P-waves, but also reflected and refracted

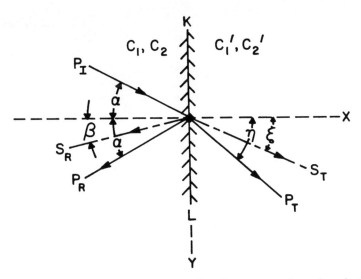

Figure 2.16. Oblique incidence of a compressional wave on a boundary between two materials having different values of C_1 and C_2.

S-waves, adds considerably to the complexity of stress wave analysis. The process by which the additional S-wave components are generated is termed "mode conversion". Similar effects occur for an incident S-wave.

For an incident P-wave, the amplitudes of the various reflected and refracted waves depend on the amplitude on the incident P-wave, the angle of incidence, the P- and S-wave velocities and the mass densities of the two media. The associated expressions, given in Rinehart (1975), are somewhat complicated and will not be included here.

In order to better appreciate the effects occurring as a result of oblique incident at a cohesive boundary, the case of a P-wave incident at a boundary between coal and sandstone, illustrated in Figure 2.17, will be examined.

Based on an assumed angle of incidence α, the following P- and S-wave velocities: $C_1 = 1100$ m/s, $C_2 = 660$ m/s, $C_1' = 2200$ m/s/ and $C_2' = 1200$ m/s , and Equation 2.22, the following relations may be derived:

$$\beta = \sin^{-1}\left[\frac{C_2}{C_1}\sin\alpha\right] = \sin^{-1}(0.6000 \sin\alpha) \qquad \text{(Eq. 2.23)}$$

$$\eta = \sin^{-1}\left[\frac{C_1'}{C_1}\sin\alpha\right] = \sin^{-1}(2.000 \sin\alpha) \qquad \text{(Eq. 2.24)}$$

$$\xi = \sin^{-1}\left[\frac{C_2'}{C_1}\sin\alpha\right] = \sin^{-1}(1.091 \sin\alpha) \qquad \text{(Eq. 2.25)}$$

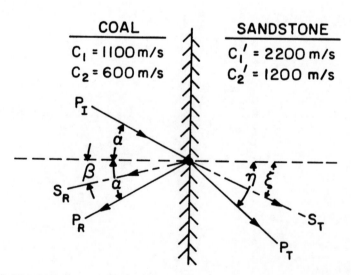

Figure 2.17. Oblique incidence of a P-wave at a boundary between coal and sandstone.

Now assuming a range of value of α between 0° and 90° the values of β, η and ξs are computed using Equations 2.23, 2.24 and 2.25, are listed in Table 2.3 and are shown plotted against α, the angle of incidence, in Figure 2.18. The data exhibits some interesting tends, namely:

1. β is approximately a linear function of α for α < ≈ 65°, for larger values of α it rapidly becomes asymptotic to a value of ≈ 37°.

2. η is approximately a linear function of α for α < 20°, for larger values of α it increases rapidly and reaches a maximum value of 90° at α = 30°.

3. ξ is approximately a linear function of α for α < ≈ 50°, for larger values of α it increases rapidly and reaches a maximum value of 90° at α ≈ 66°.

The fact that both η and ξ reach maximum values of 90° at values of α < 90° indicates that there are "critical angles" of incidence (α_c) above which the angles η and ξ are undefined (i.e., sin η > 1), that is the associated refracted P- and S-waves vanish and only the reflected P- and S-wave components remain. The critical angles may be calculated from Equation 2.22. For example,

$$\sin \alpha / C_1 = \sin \eta / C_1'$$

Table 2.3. Angles of reflection and refraction associated with oblique incidence of a P-wave at the boundary between coal and sandstone.

α	sin α	β	η	ξ
0	0	0	0	0
5	0.087	2.99	10.02	5.45
10	0.174	5.99	20.37	10.94
15	0.259	8.40	31.20	16.41
20	0.342	11.84	43.16	21.90
25	0.423	14.70	57.78	27.49
30	0.500	17.46	90.00	33.06
40	0.643	22.69	-	44.54
45	0.707	25.10	-	50.47
50	0.766	27.36	-	56.69
55	0.819	29.43	-	63.32
60	0.866	31.31	-	70.87
65	0.906	32.93	-	81.28
70	0.940	34.33	-	-
80	0.985	36.22	-	-
90	1.000	36.87	-	-

and at the critical angle α_{CP}, $\eta = 90°$, then

$$\sin \alpha_{CP} = (C_1 / C_1') \sin 90$$

$$\alpha_{CP} = \sin^{-1} (C_1 / C_1') \qquad\qquad \text{(Eq. 2.26)}$$

For the case under consideration, $C_1 = 1100$ m/s and $= 2200$ m/s, therefore

$$\alpha_{CP} = \sin^{-1} \left[\frac{1100}{2200} \right] = \sin^{-1} (0.5)$$

$$\alpha_{CP} = 30°$$

For angles of incidence $\alpha > 30°$ there will be no refracted P-wave (i.e., in Fig. 2.17, $P_T = 0$).

In a similar way the equation for computation of the critical angles α_{CS} for the refracted S-wave may be derived, namely:

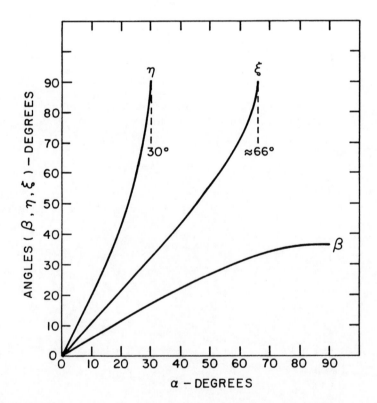

Figure 2.18. Variation of angles β, η and ξ with angle of incidence (α) for coal/sandstone interface.

$$\alpha_{CS} = \sin^{-1}(C_1 / C_2') \qquad\qquad \text{(Eq. 2.27)}$$

For the case under consideration, $C_1' = 1100$ m/s/ and C_2' m/s, therefore

$$\alpha_{CS} = \sin^{-1}(1100/1200) = \sin^{-1}(0.917)$$

$$\alpha_{CS} \approx 66.5°$$

For angles of incidence $\alpha > 66.5°$ there will be no refracted S-wave (i.e., in Fig. 2.17, $S_T = 0$).

It is clear from Equations 2.26 and 2.27 that the critical angles are functions of the wave velocities of the materials on either side of the boundary. Table 2.4 lists the variation of α_{CP} and α_{CS} for a range of values of C_1 / C_1' and C_1 / C_2'. The data are shown graphically in Figure 2.19.

The data in Table 2.4 and Figure 2.19 indicates that as the velocity contrast between the two media decreases the value of the critical angle increases. For example, the values of C_1 / C_1' $(C_1 / C_2') = 1$ both refracted P- and S-wave components will occur independent of the angle of incidence.

It should be emphasized that a stress wave traveling through a low velocity material (e.g., coal) and striking obliquely a boundary with a higher velocity material (e.g., sandstone) will behave very differently from a wave traveling through a high velocity material and striking obliquely a boundary with a lower velocity material. The former case is illustrated in Figure 2.18 and a comparison of the two cases is presented in Figure 2.20. In both cases the velocity characteristics of the coal and the sandstone are the same; however, it is clear that the angular characteristics of the various refracted components are very different. In particular it should be noted that in the former case (coal/

Table 2.4. Variation of α_{CP} (α_{CS}) as a function of C_1 / C_1' (C_1 / C_2').

C_1 / C_1' (C_1 / C_2')	α_{CP} (α_{CS})
0.01	0.57
0.05	2.87
0.10	5.74
0.20	11.54
0.30	17.46
0.40	23.58
0.50	30.00
0.60	36.87
0.70	44.43
0.80	53.13
0.90	64.16
1.00	90.00

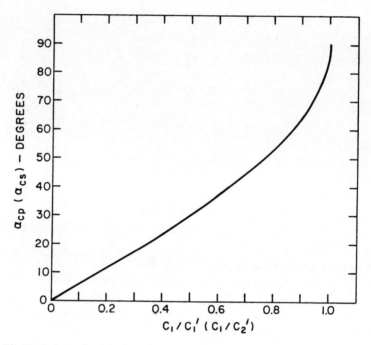

Figure 2.19. Variation of critical angle with velocity contrast.

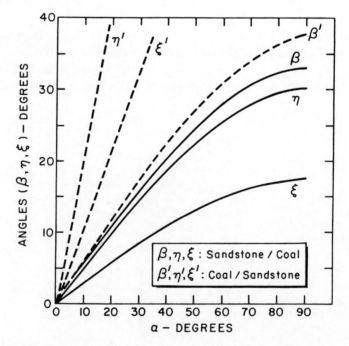

Figure 2.20. Variation of angles β, η, ξ with angle of incidence (α) for sandstone/coal and coal/sandstone interfaces.

sandstone) critical angles of incidence are associated with the refracted P- and S-waves, whereas in the later case (sandstone/coal) there are no critical angles. Furthermore it should be noted that the variation of the angle of reflection for the associated S-wave with the angle of incidence is very similar in both cases.

2.7.4 Oblique incidence-noncohesive boundary

When the boundary between two materials or between two regions of the same material is noncohesive (e.g., separated by a crack) the boundary conditions imposed on an incident stress wave are different than those for a cohesive boundary. For example, if the boundary is frictionless, stress can only be transmitted normal to the boundary. In general, however, an incident P-wave will generate both reflected and refracted P- and S-wave components. For example, Figure 2.21 illustrates relative amplitude curves for the various reflected and refracted components associated with a noncohesive boundary in materials having a Poisson's ratio (v) of 0.25 and 0.4.

It should be noted that for normal incidence ($\alpha = 0$) only a refracted P-wave occurs. As α increases the relative amplitude of the refracted P-wave initially decreases then increases, and the amplitudes of the reflected P-wave and the reflected and refracted S-wave components, which have the same amplitude, initially increase then decreases again. At $\alpha = 90°$ the relative amplitude of the refracted P-wave reaches a maximum value again and the other components drop to zero. Comparing the results for values of $v = 0.25$ and 0.40, similar

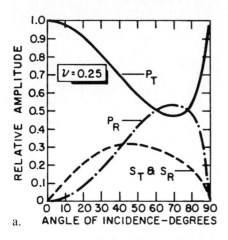
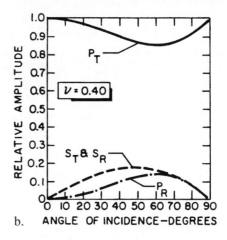

a. ANGLE OF INCIDENCE–DEGREES

b. ANGLE OF INCIDENCE–DEGREES

Figure 2.21. Relative amplitudes of reflected and refracted wave components associated with oblique incidence of a P-wave on a noncohesive boundary in materials with $v =$ 0.25 & 0.40 (after Rinehart, 1975). a. Material with $v = 0.25$. b. Material with $v =$ 0.4.

variations occur in the relative amplitudes of the various components; however, with the larger value of Poisson's ratio the variation with angle of incidence are less dramatic and the refracted P-wave is dominant.

The effects of noncohesive boundaries will certainly be of considerable importance when considering stress wave propagation in fractured rock masses. The associated theory will not be considered further here. More detailed discussion on wave propagation, and in particular energy partition at interfaces, is presented elsewhere (e.g., Egle, 1987; Rinehart, 1975).

2.8   STRESS WAVE ATTENUATION

2.8.1   General

As the stress wave from a source propagates outward through the surrounding media the energy per unit area decreases due to a number of factors, including:
1. Geometric Spreading
2. Internal Friction
3. Scattering
4. Mode conversion

As a result, the amplitude of the stress wave, which is proportional to the square root of the energy per unit area, at a specific monitoring transducer site will depend on the distance between the transducer and the source. As this distance increases the amplitude of the resulting transducer signal decreases, eventually becoming lost in the ever present background noise. Here this loss of signal amplitude will be denoted as "attenuation". Although a detailed discussion will not be presented here, a few brief comments will be included to give the reader some appreciation of the various factors involved.

Geometric spreading: As noted earlier in Section 2.2, when cylindrical and spherical waves radiate outwards from their source their energy is spread out over an increasing surface area. Since the wave amplitude is equivalent to the energy per unit area it decreases with distance from the source. For a stress wave from an AE/MS source, which is generally of the spherical type, the following relationship is valid:

$$A = A_o / r \qquad \text{(Eq. 2.28)}$$

where A_o is the stress wave amplitude at the source, and A is the amplitude at a distance r from the source. It should be noted that this factor depends only on geometry and is independent of the properties of the materials involved.

Internal friction: Since the propagation of a stress wave through a medium involves a dynamic strain response any anelastic characteristics of the medium will result in the loss or absorption of energy. Such effects as hysterisis and viscoelastic dampling cause internal friction and resulting energy loss. In general such loss is found to be exponential with distance, namely:

$$A = A_0 e^{-\alpha r} \qquad \text{(Eq. 2.29)}$$

where α is often referred to as the attenuation coefficient, and the other factors are defined earlier. Past research has indicated that α is dependent on stress wave frequency, however, the exact form of the dependency is uncertain.

Scattering: In a polycrystalline material (e.,g., rocks) a phenomenon known as scattering occurs when the wavelength of the stress wave becomes comparable with the grain size, namely:

$$d \approx \lambda$$

$$\lambda = C.f \qquad \text{(Eq. 2.30)}$$

where d is the mean grain diameter, and λ, C and f are, respectively, the wavelength, propagation velocity and frequency of the incident stress wave. Under the conditions that $d \approx \lambda$ the incident stress wave sees the grain as an obstacle resulting in the generation of secondary waves which spread out in a variety of directions reducing the general outward flow of energy. Attenuation due to scattering may be important in some laboratory situations where high frequency signals are involved. In most field situations, where low frequencies are normally involved, scattering effects from such features as rock grains are unimportant. However, in blocky rock masses, where the block dimensions are of the order of the wavelength of the incident stress wave, scattering may play a role.

Mode conversion: As noted earlier in Section 2.7, a stress wave incident on a cohesive or noncohesive boundary causes the generation of a variety of refracted and reflected components. This effect, termed mode conversion, effectively reduces the level of outgoing energy and will thus contribute to the reduction of stress wave amplitude at a remote transducer site.

2.8.2   Typical attenuation characteristics

In geologic materials, particularly in field situations, little information is available in regard to the specific role played by the various factors discussed in the preceding section. Some authors (e.g., Dobrin, 1960), however, indicate that an expression based on a combination of Equations 2.28 and 2.29 is useful, namely:

$$A = A_0 \, (e^{-\alpha r} / r) \qquad \text{(Eq. 2.31)}$$

where α is considered to be a function of frequency and all factors have been defined earlier.

Overall attenuation factors for use in most AE/MS studies are obtained by locating two transducers on the test specimen or field structure under study, activating a suitable artificial source, monitoring the amplitude of the resulting stress wave at the two transducer sites, and using the following equation to compute an equivalent attenuation factor (α_e), namely:

$$\alpha_e = \frac{20}{d} \log \frac{A_1}{A_2} \qquad \text{(Eq. 2.32)}$$

where α_e is the equivalent attenuation factor, d is the distance between transducers, and A_1 and A_2 are the peak signal amplitudes at transducer 1 and 2. If d is in meters, α_e is in dB/m.

Figure 2.22 illustrates values of α_e obtained for soils, rocks and coal. Values for iron and steel are included for comparison. It is clear that the results are highly sensitive to frequency, with values of α_e increasing with increasing frequency. For example, for rock and coal, values of α_e at low frequencies (e.g., f ≈ 100 Hz) are of the order of 10^{-2} to 10^{-3} dB/m. At higher frequencies (e.g., f ≈ 300 kHz) values of α_e are the order of 10^1 to 10^3 dB/m.

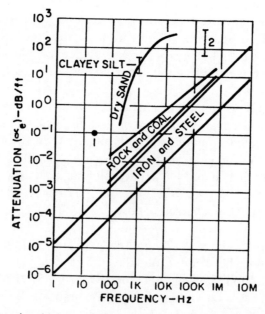

Figure 2.22. Attenuation (α_e) as a function of frequency for various materials. (Major data from Koerner et al. (1981); additional data: 1 – rock/soil structure (Hardy, 1979), 2 – sedimentary rock types (Hardy, 1974); conversion factor, 1.0 dB/ft = 3.28 dB/m.)

2.9 DETECTION OF AE/MS SIGNALS

2.9.1 General

As discussed earlier in this chapter an AE/MS source generates stress waves which propagate outward from the source. After traversing the surrounding, often complex, media they eventually reach one or more of the transducers installed to monitor the AE/MS activity. Such transducers respond to the particle motion produced by the incident stress waves and generate equivalent electrical signals. These signals are subsequently processed to provide the required experimental results. Figure 2.23 illustrates the flow of information through a generalized AE/MS monitoring system.

In most cases the transducer is the primary interface between the investigator and the specimen or structure under study. Both theory and practice have indicated that this interface is probably the most critical point in the overall AE/MS monitoring system. Once specific components of the mechanical energy in the stress wave has been converted to an electrical signal these are the only components available for signal conditioning and subsequent analysis. As a result, the transducer type, orientation and location, and the suitability and quality of the associated installation is extremely important.

This section will deal specifically with the detection of particle motion, the general character of the detected signals, determination of the propagation direction, and optimization of transducer orientation. Specific details in regard to the design of various types of AE/MS monitoring transducers and associated installation techniques will be considered later in Chapters 3 and 4.

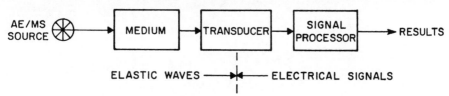

Figure 2.23. Information flow through a generalized AE/MS monitoring system.

2.9.2 Detection of particle motion

Assume, as illustrated in Figure 2.24, that an AE/MS source is activated at a point (S) in a homogeneous, isotropic solid, and an associated stress wave propagates outward from the source. The mechanical disturbance (particle motion) due to the stress wave at the point P may be monitored in terms of displacement (s), velocity (v) or acceleration (a). Suitable transducers may therefore be a displacement gage, velocity gage (geophone) or an accelerometer. The type of transducer used will depend on the frequency content of the associated particle motion.

Assuming, as shown in Figure 2.25, that for a simple analysis the particle motion may be considered to be sinusoidal in form, the displacement may be written as,

$$S = S_o \cos \omega t \qquad\qquad (Eq.\ 2.33)$$

Now velocity and acceleration may be determined, by suitable differentiation of Equation 2.33, namely:

$$v = ds/dt = -\omega s_o \sin \omega t \qquad\qquad (Eq.\ 2.34)$$

and

$$A = dv/dt = -\omega^2 s_o \cos \omega t \qquad\qquad (Eq.\ 2.35)$$

Maximum values for s, v and a, neglecting signs, will be as follows:

$$\left.\begin{array}{l} s_m = s_o \\ v_m = \omega s_o \\ a_m = \omega^2 s_o \end{array}\right\} \qquad\qquad (Eq.\ 2.36)$$

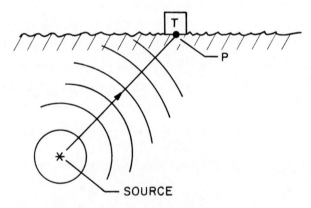

Figure 2.24. Transducer monitoring particle motion at point P due to stress wave emitted from source location.

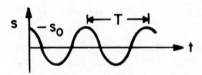

Figure 2.25. Assumed particle displacement as a function of time.

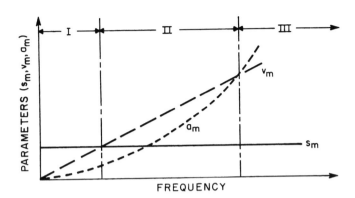

Figure 2.26. Variation of Particle Motion Parameters (s_m, v_m, a_m) as a Function of Frequency Assuming Sinusoidal Particle Displacement.

A general plot of these parameters as a function of frequency ($\omega = 2\pi f$) is shown in Figure 2.26.

Based on the general trends shown in Figure 2.26, it is clear that in certain frequency regions it is desirable to utilize specific types of transducers. For example in region I, low frequencies (e.g., 0.01-1.0 Hz), displacement gages are most efficient; in region III, high frequencies (e.g., 5 kHz – 500 kHz), accelerometers are most efficient; and velocity gages (geophones) are most efficient in region II, the intermediate frequency range (e.g., 1.0 Hz – 5 kHz). It should be noted, however, that the basic sensitivity versus frequency characteristics of the three transducer types must also be considered in selection of the optimum type. Further details in regard to selection of transducers are presented later in Chapter 3 and 4.

2.9.3 Computation of propagation direction

When monitoring particle motion it is important to consider the direction of propagation of the associated stress wave relative to the axis of sensitivity of the transducer. Since the line joining the source and the monitoring transducer may be considered a vector, the necessary vector concepts required for computation of propagation direction will be briefly reviewed in this section.

Figure 2.27 illustrates the positions of an AE/MS source (S) and a monitoring transducer (P). The straight line joining the points S and P may be described as the vector, $\vec{A}$. Following standard vector notation, as shown in Figure 2.27, $\vec{A}$ may be considered to consist of three components A_x, A_y, and A_z parallel to the x-, y- and z-axis, respectively.

Now a vector $\vec{A}$ may be specified by its magnitude A and its direction cosines, as defined in Figure 2.27b, namely:

$$\left.\begin{array}{l} a_x = \cos \alpha \\ a_y = \cos \beta \\ a_z = \cos \gamma \end{array}\right\} \qquad \text{(Eq. 2.37)}$$

The components of the vector may be computed from the following relationships:

$$\left.\begin{array}{l} A_x = A \cos \alpha = A\, a_x \\ A_y = A \cos \beta = A\, a_y \\ A_z = A \cos \gamma = A\, a_z \end{array}\right\} \qquad \text{(Eq. 2.38)}$$

where

$$A = \sqrt{(A_x)^2 + (A_y)^2 + (A_z)^2} \qquad \text{(Eq. 2.39)}$$

Furthermore, the direction cosines may be computed from the following relationships:

$$\left.\begin{array}{l} a_x = A_x / A \\ a_y = A_y / A \\ a_z = A_z / A \end{array}\right\} \qquad \text{(Eq. 2.40)}$$

where

$$(a_x)^2 + (a_y)^2 + (a_z)^2 = 1 \qquad \text{(Eq. 2.41)}$$

As an example consider a source located at a point S (-500, 300, 300) and a monitoring transducer at the point P (0, 1000, 1000). Now

$$A_x = 500$$
$$A_x = 700$$
$$A_z = 700$$

and from Equation 2.39,

$$A = \sqrt{(A_x)^2 + (A_y)^2 + (A_z)^2} = \sqrt{(500)^2 + (700)^2 + (700)^2}$$

giving

$$A = 1,109$$

From Equation 2.40 the direction cosines are as follows:

$$a_x = \frac{A_x}{A} = \frac{500}{1,109} = 0.4509$$

$$a_y = \frac{A_y}{A} = \frac{700}{1,109} = 0.6312$$

$$a_z = \frac{A_z}{A} = \frac{700}{1,109} = 0.6312$$

and $(a_x)^2 + (a_y)^2 + (a_z)^2 = 0.2033 + 0.3984 + 0.3984 = 1.0$ as required. Finally the related angles from Equation 2.37 are as follows:

$$\alpha = \cos^{-1}a_x = 63.2°$$
$$\beta = \cos^{-1}a_y = 50.9°$$
$$\gamma = \cos^{-1}a_z = 50.9°$$

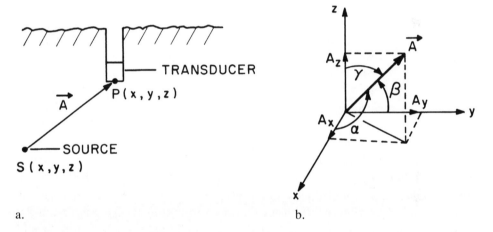

a.   b.

Figure 2.27. Vector relationship between source and monitoring transducer. a. Relative position of source and transducer. b. Vector components.

2.9.4 Character of detected signal

Besides the losses in signal amplitude due to various types of attenuation, the general character of an AE/MS signal changes with distance from the source. Reasons for the latter include such factors as the presence of multiple wave components, reflection at boundaries and interfaces, and dispersion.

Multiple wave components: In general a typical AE/MS source will generate a stress wave containing both P- and S-wave components. Even if the source produced a pure P- or S-wave, as a result of mode conversion during propagation, the stress wave reaching the monitoring transducer will in most cases contain both P- and S-wave components. For example, Figure 2.28 illustrates a typical AE/MS signal detected by a down-hole accelerometer located some 1800 m from an underground AE/MS source. Due to the relatively large source-to-

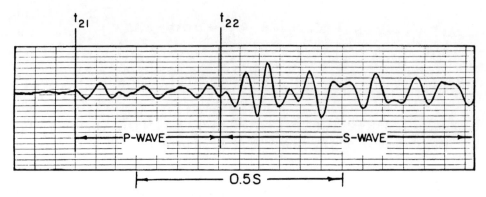

Figure 2.28. Typical AE/MS signal detected at a relatively large distance from the associated source. (Both P- and S-wave components are visible).

transducer distance, and the fact that the P- and S-wave velocities are different, both the P- and S-wave components are clearly visible. As noted these arrive at time t_{21} and t_{22}, respectively.

Although the transducer which detected the signal shown in Figure 2.28 was designed specifically for detection of P-waves, due to a number of factors, including the propagation direction relative to the axis of sensitivity of the transducer, the transducer transverse sensitivity and mode conversion effects, the detected signal will normally contain both P- and S-wave components. However, when the monitoring transducer is located relatively near the source the components are superimposed and the presence of the two components is not clearly evident. Figure 2.29, for example, shows the relative position of the P- and S-wave components of the signal illustrated in Figure 2.28 as they would appear at the source (S), and at two transducers (T1 and T2) located at distances d_1 and d_2 from the source. At each position the actual signal detected by the transducer would be the superposition of the electrical signals due to both the P- and S-wave components. As indicated, for increasing distances from the source the S-wave, due to its lower velocity, falls progressively further behind the P-wave. At d_2 the two components just begin to separate and the superimposed signal would appear as shown in Figure 2.28.

When obvious separation of the P- and S-wave components can be observed in the detected signal, such signals can be utilized to calculate the transducer-to-source distance. Such calculation is based on straight line propagation between the source and the transducer. As illustrated in Figure 2.29, transducer T2 is located at an unknown distance d_2 from the source and the P- and S-wave components arrive at time t_{21} and t_{22}, respectively. For the P-wave,

$$t_{21} = d_2/C_1$$

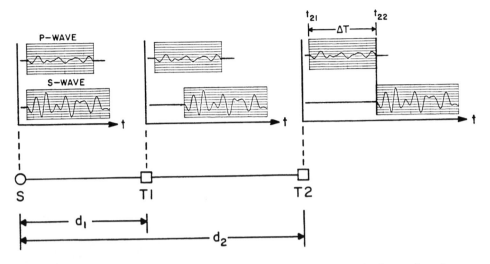

Figure 2.29. Relative position of P- and S-wave signals at an AE/MS source and at two transducer sites some distance away.

and for the S-wave,

$$t_{22} = d_2/C_2$$

where C_1 and C_2 are the associated P- and S-wave velocities.

Now

$$\Delta t = t_{22} - t_{21}$$

and substituting for t_{21} and t_{22}, and rearranging, the following expression is obtained for d_2,

$$d_2 = \left[\frac{C_1 C_2}{C_1 C_2} \right] \Delta t \qquad \text{(Eq. 2.42)}$$

Now Δt may be obtained from the signal waveform, and if C_1 and C_2 are known d_2 may be calculated. If only C_1 is known the generally accepted relationship,

$$C_2 \approx 0.6 \, C_1$$

may be utilized and Equation 2.42 reduced to the following:

$$d_2 = (1.5 C_1) \, \Delta t \qquad \text{(Eq. 2.43)}$$

For example, considering the waveform shown earlier in Figure 2.28,

$$\Delta t \approx 0.35 \text{ s}$$

and assuming

$$C_1 \approx 3500 \text{ m/s}$$

Then

$$d_2 = 1.5 \times 3500 \times 0.35$$

$$d_2 \approx 1837 \text{ m}$$

Reflection at boundaries and interfaces: The overall particle motion detected by the transducer includes the components due to the direct signals as well as secondary components due to reflections from boundaries and interfaces. Since the front edge of the direct signal always arrives first, the initial section of the detected p-wave signal will be free of secondary components. Subsequent regions of the signal, however, may become more and more contaminated by secondary components. These components are difficult to separate using single component transducers.

Dispersion: This is a phenomena caused by the frequency dependence of wave speed in certain physical systems. In general AE/MS waves exhibit dispersion when propagating in a solid medium in which the wavelength is comparable to one or more of the dimensions of the medium. Because the wave speed is a function of frequency, a disturbance, made up of various frequency components, will change in form as the disturbance propagates since each of the components will propagate at a different speed. Dispersion effects are particularly important in structures such as thin plates and thin-walled vessels.

2.9.5   Transducer signal components

The majority of the commonly utilized AE/MS transducers have a specific axis along which they are most sensitive, i.e., particle motion along this axis generates the maximum electrical output. To date, however, in both field and laboratory studies relatively little consideration has been given to the problem of optimum transducer orientation. In general, for maximum sensitivity the transducer axis should be oriented parallel to the direction of stress wave propagation for detection of P-waves and perpendicular to this direction for detection of S-waves. In the majority of cases monitoring transducers are mounted vertically. As a result their electrical output will depend on the direction of propagation of the incident stress wave, and will normally contain both P- and S-wave components (e.g., see Fig. 2.28).

For example, consider a pure P-wave, P(t) propagating in the direction A-B, at an angle γ to the vertical, as shown in Figure 2.30a. At the point M the P-wave may be resolved into three components, namely:

$$
\left.
\begin{aligned}
P_x\,(t) &= a_x\,P(t) \\
P_y\,(t) &= a_y\,P(t) \\
P_z\,(t) &= a_z\,P(t)
\end{aligned}
\right\}
\qquad\text{(Eq. 2.44)}
$$

where a_x, a_y, and a_z are the associated direction cosines of the propagation direction.

Use of a single vertical transducer: Based on Equation 2.44, a single vertically oriented monitoring transducer located at point M, therefore, will have an electrical output proportional to $P_z\,(t) = a_z\,P(t)$, where $0 < a_x < 1$ depending on the propagation direction. In general, from Equation 2.37, $a_z = \cos\gamma$ and Figure 2.31 illustrates the variation of a_z as a function of γ. It is clear that the effective transducer output drops rapidly as the angle of propagation (γ) increases.

The use of a single vertical transducer also effectively limits the area over which AE/MS monitoring of P-waves can be carried out. Figure 2.32 illustrates a simple field situation where a surface mounted transducer is used to monitor subsurface AE/MS activity due to a source (S). The relationship between the depth (d) of the structure being monitored, the horizontal distance (R) from the transducer to the source (S) and the angle γ is as follows:

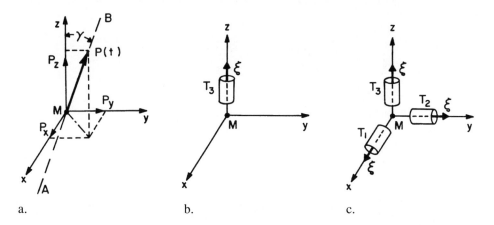

a.   b.   c.

Figure 2.30. Components of a propagating P-wave and various monitoring transducer configurations (ξ denotes the axis of maximum sensitivity of the transducer). a. P-wave propagating in direction A-B and associated components. b. Single vertical transducer. c. Triaxial transducer.

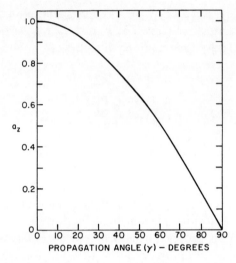

Figure 2.31. Variation of direction cosine (a_z) with propagation angle (γ).

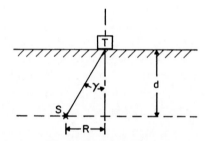

Figure 2.32. Geometrical relationship for an underground source (S) and a surface mounted transducer (T).

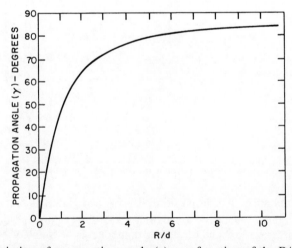

Figure 2.33. Variation of propagation angle (γ) as a function of the R/d ratio.

$$R = d \tan \gamma$$

or rewriting

$$R/d = \tan \gamma \qquad \text{(Eq. 2.45)}$$

Figure 2.33 illustrates the variation of γ as a function of the R/d ratio. As R/d increases the value of γ increases, and, as shown earlier in Figure 2.31, this results in a decreases in the value of a_z and a loss of effective transducer sensitivity for P-wave detection. However, for a specific value of γ, since the R/d ratio is inversely proportional to depth (d), as the depth of the monitored structural increases the associated horizontal transducer-to-source distance will also increase.

Combining Equations 2.45 and 2.37 an equation relating a_z, d and R may be obtained, namely:

$$a_z = \frac{d}{\sqrt{R^2 + d^2}} \qquad \text{(Eq. 2.46)}$$

For example, for a monitored structure located 20 m below surface and a horizontal source to transducer distance of 10 m,

$$a_z = \frac{20}{\sqrt{(10)^2 + (20)^2}} = 0.894$$

Equation 2.46 may also be rewritten in the following form:

$$R = (d\sqrt{1-a_z^2}) / a_z \qquad \text{(Eq. 2.47)}$$

which is useful to calculate the horizontal transducer-to-source distance for a specific structure depth and desired value of a_z. For example for a monitored value of A_z of 0.8,

$$R = (200\sqrt{1-(0.8)^2}) / 0.8 = 150 \text{ m}$$

Use of a triaxial transducer: Based on the preceding discussions it is apparent that although a single vertically oriented transducer may be used to monitor P-wave signals, the magnitude of the detected signal is highly dependent on the propagation angle (γ), i.e., the location of the source relative to the transducer. A more suitable arrangement would be to utilize a triaxial transducer unit, as illustrated earlier in Figure 2.30c. Such a transducer unit contains three separate monitoring transducers (T1, T2 and T3) oriented at 90° to one another. As noted earlier in Equation 2.44 this arrangement will allow all three components of the P-wave to be detected, namely:

$$P_x (t) = a_x P(t)$$
$$P_y (t) = a_y P(t)$$
$$P_z (t) = a_z P(t)$$

and the resultant, i.e. the true magnitude of the P-wave [P(t)], computed using Equation 2.39, namely:

$$P_t = \sqrt{(P_x)^2 + (P_y)^2 + (P_z)^2} \qquad \text{(Eq. 2.48)}$$

Furthermore using Equation 2.40 it is also possible to obtain the direction cosines of the line defining the propagation direction, namely:

$$\left. \begin{array}{l} a_x = P_x (t) / P(t) \\ a_y = P_y (t) / P(t) \\ a_z = P_z (t) / P(t) \end{array} \right\} \qquad \text{(Eq. 2.49)}$$

These direction cosines along with the transducer-to-source distance determined earlier using Equation 2.42 is the basis of the hologram method of source location which will be discussed further in Chapter 5.

Unfortunately, in many cases, noise signals, due to a variety of sources, are also present on the monitoring channels, namely:

$$P'_x (t) = a_x P (t) + e_x (t)$$
$$P'_y (t) = a_y P (t) + e_y (t) \qquad \text{(Eq. 2.50)}$$
$$P'_z (t) = a_z P (t) + e_z (t)$$

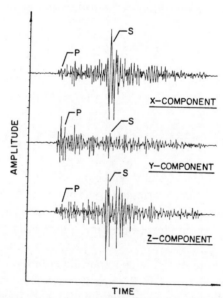

Figure 2.34. AE/MS signals detected using a triaxial monitoring transducer.

where $P'_x(t)$, $P'_y(t)$ and $P'_z(t)$ are the recorded x-, y- and z-axis signals and $e_x(t)$, $e_y(t)$ and $e_z(t)$ are the associated noise signals.

Figure 2.34 illustrates a set of typical x-, y- and z-components of a stress wave generated by an underground AE/MS source and monitored using a triaxial transducer. Each of the signals exhibit P- and S-wave components plus noise. It should be noted that although all three signals exhibit P- and S-wave components the degree to which each component is present is dependent on the orientation of the associated transducer. For example the x- and z-component shows a large S-wave and a small P-wave component. In contrast the y-component shows a large P-wave and a small S-wave component. Computer analysis of the 3-component data makes it possible to determine the true values of the P- and S-wave signals.

REFERENCES

Borm, G.W. 1978. Methods from Exploration Seismology: Reflection, Refraction and Borehole Prospecting, *Proceedings Symposium on Dynamical Methods in Soil and Rock Mechanics*, Vol. 3, Rock Dynamics and Geophysical Aspects, Balkema, Rotterdam, pp. 87-114.

Breckenridge, F.R., Proctor, T.M., Hsu, N.N., Fick, S.E. and Eitgen, D.G. 1990. Transient Sources for Acoustic Emission Work, *Progress in Acoustic Emission VI*, Proceedings, 10th International Acoustic Emission Symposium, Sendai 1990, Japanese Society for Non-Destructive Inspection, Tokyo, pp. 20-37.

Dobrin, M.B. 1960. *Introduction to Geophysical Prospecting*, McGraw-Hill Book Company, New York.

Dowding, C.H. 1985. *Blast Vibration Monitoring and Control*, Prentice-Hall Inc., Englewood Cliffs, N.J., pp. 6-23.

Dresen, L. and Ruter, H. 1995. *Seismic Coal Exploration, Part B: In-Seam Seismics*, Pergamon Press, Elsevier Science Inc., New York, pp. 15-93.

Egle, D. 1987. Wave Propagation, *Nondestructive Testing Handbook*, Vol. 5: Acoustic Emission Testing, Section 5 (ed. P.M. McIntire), American Society for Nondestructive Testing, Columbus, Ohio, pp. 92-120.

Ewing, W.M., Jardetzky, W.S. and Press, F. 1957. *Elastic Waves in Layered Media*, McGraw-Hill, New York.

Gibowicz, S.J. and Kijko, A. 1994. *An Introduction to Mining Seismology*, Academic Press, New York, pp. 24-47.

Hardy, H.R. Jr. 1974. *Microseismic Techniques Applied to Coal Mine Safety*, USBM Grant No. G0101743 (Min-45), The Pennsylvania State University, USBM Open File #23-75, NTIS, PB 241-208/AS.

Hardy, H.R. Jr. 1979. *Microseismic Study at the New Entrance Area, Mammoth Cave National Park, Mammoth Cave, Kentucky*, Confidential Report, CR-HRH/79-1.

Koerner, R.M., McCabe, W.M. and Lord, A.E. Jr. 1981. "Acoustic Emission Behavior and Monitoring of Soils", *Acoustic Emissions in Geotechnical Engineering Practice*, V.P. Drnevich and R.E. Gray (eds.), STP 750, American Society for Testing and Materials, Philadelphia, Pennsylvania, pp. 93-141.

Kolsky, H. 1963. *Stress Waves in Solids*, Dover Publications, New York.

Paillet, F.L. and Hon Cheng, C. 1991. *Acoustic Waves in Boreholes*, CRC Press, Boston, pp. 37-59.
Pierce, A.D. 1981. *Acoustics: An Introduction to Its Physical Principles and Applications*, McGraw-Hill Book Company, New York.
Rinehart, J.S., 1975. *Stress Transients in Solids*, Hyperdynamics, Santa Fe.

CHAPTER 3

Field monitoring systems

3.1 INTRODUCTION

Chapters 3 and 4 will include a general discussion on the form and overall design of field and laboratory AE/MS monitoring systems. In general, as indicated in Figure 3.1, a single channel monitoring system can be considered to consist of three major sections, namely: a transducer, a signal conditioning section and a readout section. The general characteristics of each of the three sections will be briefly considered. A more detailed discussion of certain aspects (e.g., transducers for use in special field applications, special signal conditioning techniques, remote monitoring, etc.) will be included later in the text.

It is important to reiterate that the major differences between AE/MS field and laboratory studies is one of scale and frequency content (see figure 1.2 presented earlier). For example, laboratory studies often involve only a single transducer, short transducer-to-source distances (e.g., 0.01-2 m), and the detection and processing of AE/MS signals containing relatively high frequency components (e.g., 50 kHz – 500 kHz). In contrast, field studies usually involve a multi-transducer array, array dimensions and transducer-to-source distances often of the order of several hundreds to thousands of meters, and stress wave signals containing relatively low frequency components (e.g., 1 Hz – 5 kHz).

Due to the primary importance of field oriented AE/MS application in geomechanics, consideration will therefore be given initially to field

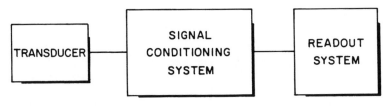

Figure 3.1. Block diagram of a single channel AE/MS monitoring system.

monitoring systems. Later in Chapter 4, monitoring system design and operating characteristics particularly applicable to laboratory scale AE/MS studies will be considered. Additional references relative to field AE/MS transducers and monitoring systems are presented in other publications, including: Dowding (1985), Dresen and Ruter (1995, and Galperin et al. (1986).

3.2   GENERAL CONSIDERATIONS

The design and development of a suitable AE/MS field monitoring system necessitates consideration of the overall system, which includes the field site itself, the transducers (electro-mechanical characteristics, installation technique, and array geometry), and the monitoring facilities.

It is important to realize that, at the input to the transducer, the mechanical signals associated with AE/MS activity are often of very low amplitude (e.g., $v \approx 10^{-7}$ m/s in some cases) and extremely high-gain systems are often required to convert these to useable electronic signal levels (normally 1-10 volts). Since system gains of x 100,000 (100 dB*) are common, great care in the design, construction, and maintenance of such monitoring systems are required. Furthermore, since only very limited data can be obtained using a single transducer (single channel system), most field systems utilize a number of transducers (an array) and hence the associated monitoring facilities involves a number of parallel channels (usually a minimum of five) and an associated multi-channel readout.

3.3   SYSTEM OPTIMIZATION

During the last 25 years there has been considerable development in the application of AE/MS techniques to monitoring stability of underground and surface structures. As research in this area continues to grow, it has become more and more evident that the overall monitoring system necessary for each field project is "site-specific". As a result one of the most critical phases in the initiation of a new AE/MS field monitoring project involves the selection of the most suitable instrumentation, and the optimization of this instrumentation for the specific site conditions.

The development of a suitable overall monitoring system necessitates the following:

1. Selection of the most suitable transducer(s).

---

* The voltage gain (or loss) in an electronic unit such as an amplifier is often expressed in terms of decibels (dB), where dB = 20 $\log_{10}$ (E_o/E_{in}) and E_{in} and E_o are the input and output voltages. The ratio (E_o/E_{in}) = G is commonly defined as the amplification factor.

2. Development of the most efficient transducer installation technique(s) and array geometry.
3. Careful attention to the method(s) by which AE/MS signals are coupled from the various transducer locations to the monitoring facilities.
4. Design of a stable and sensitive monitoring facility.
5. Detailed preliminary field studies to evaluate the ambient field site conditions so that the overall monitoring system can be fully optimized.

Only after the preceding steps have been taken is it realistic to expect that the overall monitoring system will provide meaningful data.

3.4   CLASSES OF MONITORING

As has been pointed out earlier, AE/MS activity at a field site is monitored by installing a number of transducers (an array), in locations where they can detect any activity which may be originating in the structure under study. In general, there are two classes of monitoring, depending on the aims of the study and the number of transducers involved, namely: "general monitoring" and "location monitoring".

3.4.1   General monitoring

The aim of a general field monitoring system is to establish if AE/MS activity is being generated in the general area of the structure under investigation, and to ascertain if this activity is associated with changes in various other monitored field parameters such as storage pressure level, rate of tunnel advance, pillar loading, roof sag, etc. Figure 3.2, for example, illustrates a single transducer general monitoring system installed in a tunnel roof to evaluate the degree of overbreak occurring prior to installation of permanent support. Since such monitoring involves only a single transducer, sufficient arrival-time data is not available to determine the actual source of the AE/MS activity.* Furthermore, it is not possible to derive unique location information from the magnitude of the observed AE/MS events since signals of similar amplitude can be obtained from a small near-event as from a large distant-event.

The use of general monitoring is further complicated by the fact that there may be possible sources of AE/MS and background activity at a specific field site which are unrelated the structure under investigation. These include electrical transients, vehicular traffic, blasting, low-level seismic activity, etc.

---

* As discussed earlier, P-wave / S-wave separation can in some cases be used to obtain approximate transducer-source distances. This is normally only possible when such distances are relatively large (e.g., 1,000 m).

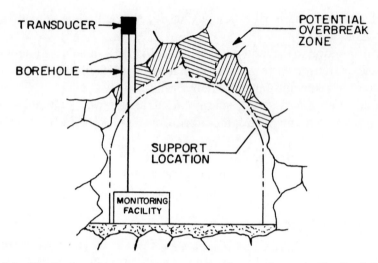

Figure 3.2. Single transducer arrangement for "general monitoring" of degree of overbreak in a tunnel prior to support installation.

Unless activity from such sources can be effectively eliminated using selective electronic filtering of the associated signals (which is often very difficult and in some cases impossible), by limiting the range of the system (spatial filtering), or by recording only during specific periods of time, the use of general monitoring may be impractical.

3.4.2   Location monitoring

As has been pointed out in the previous section, general monitoring has limited application in most situations, and it is necessary to locate the source of all AE/MS events in order to insure that only AE/MS data known to be associated with the structure under study are utilized. Source location of individual events is only possible if data is available from a suitable array of transducers. A minimum of four transducers are required for epicenter location (location in the horizontal plane), with source depth undetermined; and a minimum of five transducers are required for three-dimensional, or hypocenter location.

3.5   FIELD TRANSDUCERS

3.5.1   Transducer types

The purpose of an AE/MS transducer is to convert the mechanical energy associated with an AE/MS event into a suitable electrical signal. When a geologic structure is loaded, mechanical signals are generated due to localized deformation and/or failure in areas of high stress concentration. As noted earlier, AE/MS activity at a specific point in the structure may be detected by monitoring the particle displacements, velocities or accelerations generated by the associated stress waves at that point using a suitable transducer. Where signals containing relatively high frequency components (f > 5 kHz) are involved, accelerometers are usually employed. In contrast, low frequency signals (f < 1 Hz) are usually detected with displacement gages. Signals between these extremes (1 Hz < f < 5 kHz) are conveniently detected using velocity gages. Normally geophones (velocity gages) or accelerometers are used in AE/MS field studies and these have sensitivities in the range 40-400 V/m/s and 2-100 mV/g, respectively. Displacement gages are rarely used in AE/MS studies except in very special applications. Figure 3.3 illustrates a number of commercially available AE/MS transducers suitable for field applications.

In general, the sensitivity of all three types of transducers is a maximum for mechanical signals propagating in a direction parallel to the axis of the associated transducer sensing element. It is important to note that although the sensitivity of most accelerometers are independent of the mounting angle, such is not the case for velocity and displacement transducers. The latter types are often available in two models for either vertical or horizontal mounting, and these must be installed as close to the specified mounting angle as possible to obtain maximum sensitivity. Further, since the vertical models are designed to "hang" vertically in the gravity field, when mounting such units on the roof of a structure (e.g., adhesive mounting shown later in Figure 3.6) care must be taken to reverse the direction of the sensing element.

Although both geophones and accelerometers have been used in monitoring such geologic structures as mines and tunnels, in the majority of cases, geophones are employed. These transducers are less expensive than accelerometers and are generally more sensitive in the lower frequency range where much of the AE/MS energy associated with instabilities in large structures normally occur. A brief description, relative to the basic principles and operating characteristics of geophones will be presented later in this section. Further details relative to the construction and operating characteristics of accelerometers are presented later in Chapter 4.

Geophone design varies from manufacturer to manufacturer; however, Figure 3.4 illustrates the basic concept of the moving coil type geophone (velocity gage) used in the majority of AE/MS field studies. In this type of geophone, a mass is supported by a spring from an associated support which is mounted on the structure under study. A wire (or coil) is attached to the mass, and suspended in the field of a permanent magnet. When the structure moves, the magnet and support also move. The mass, however, tends initially to remain stationary and lags behind the motion of the structure resulting in a relative motion between the coil and magnetic field. The resultant voltage output is proportional to the velocity of this motion, as a result, such a devise is termed

a.   b.

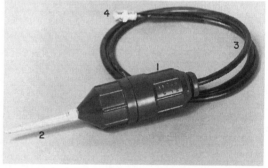

d.

c.

Figure 3.3. Typical commercially available AE/MS transducers suitable for field applications. a. Typical accelerometer units. b. Basic geophone sensing element. c. Low frequency geophone unit. d. Typical marsh-type geophone unit. (1. marsh case, 2. spike, 3. electrical cable, and 4. electrical connector).

a velocity gage. Now, if the structure suddenly moves upward (or downward) and remains there, and the magnet and support follow perfectly, the mass will lag, return to its original position relative to the magnet, and continue moving in a manner similar to a vertical pendulum. This oscillation will gradually die out due to the energy required to move the air, and losses in the spring and in the magnetic circuit. If the coil is loaded or shorted, a current is caused to flow in

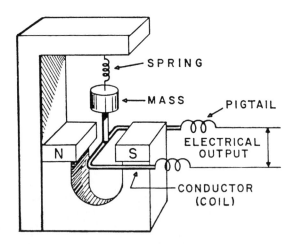

Figure 3.4. Simple model illustrating the basic concept of the moving coil type geophone (after Anon., 1984).

the coil which produces a force that opposes the motion. By properly designing a geophone, this force can be adjusted to cause the oscillation to die out in any specified time. When the geophone is critically damped (i.e., a suitable external shunt or damping resistor is connected across the coil), the mass never completely returns to its relative position with the magnet and there is no overshoot or oscillation. This is called viscous damping due to its similarity to the resistance to motion caused by a fluid or gas (i.e., viscosity).

Figure 3.5 illustrates the frequency response of a typical geophone, Geospace, type GSC-11D having a natural frequency of 14 Hz. Here the logarithm of the output voltage is shown plotted against the logarithm of the driving frequency. Curves are presented for three different shunt resistor values, open, 910 Ω and 360 Ω, with resulting damping of 23%, 50% and 70%.

The natural frequency of the geophone is defined as the frequency of oscillation in the theoretical case of no damping. This frequency is controlled by the ratio of the total mass to the spring constant. The geophone output (or transduction constant) is related to the product of the wire length of the coil and the average magnetic flux density at right angles to the coil. A number of commercial firms which manufacture geophones suitable for use in AE/MS field studies are listed in the Appendix.

A limited number of field studies have also been conducted both by the author and others (e.g., Batchelor et al., 1983; Meister, 1980; Power, 1977; Schulte, 1980; Shuck and Keech, 1977), using hydrophones as AE/MS transducers. Such transducers are effectively ultra-sensitive pressure transducers (see Doebelin, 1975) and when installed in a fluid-filled borehole they sense the presence of minute pressure changes in the borehold fluid

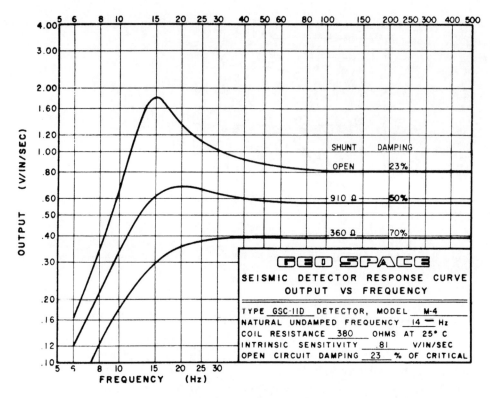

Figure 3.5. Frequency response curves for geospace type GSC-11D, model M-4, 14 Hz geophones. (Curves for three shunt resistance values and equivalent percent damping are shown. Scale factor: 1.0 v/in./sec = 39 v/m/sec.) (Courtesy OYO Geospace Co., Houston, Texas).

resulting from mechanical instabilities occurring in the surrounding strata. While hydrophones have proven extremely successful in other fields, such as underwater detection, they present a number of special problems in the geotechnical area. For example, fluid-filled boreholes are generally required and AE/MS source location may be difficult due to the composite media (fluid/rock) involved. It should be noted, however, that although hydrophones are not applicable as AE/MS transducers in most geotechnical field studies, they have proven extremely useful in a number of special applications.

Regardless of the type of transducer utilized, it is important that where possible such transducers be suitably calibrated. This is particularly critical when transducers are reused at a number of field sites. A computer-based system for calibration of velocity gages (geophones) and accelerometers have recently been developed in the Penn State Rock Mechanics Laboratory (Oh, 1998). Further discussion on transducer calibration will be included later in Chapter 6.

3.5.2 Transducer location

Figure 3.6 illustrates a simplified field situation where it is desired to monitor AE/MS activity associated with an underground geologic structure (ST). This structure, which could be, for example, a tunnel, underground mine, or a storage area, is assumed to be located at an average depth below surface of d_1. It is also assumed that the site is overlain with a layer of unconsolidated material, for example, soil, to a depth of d_2. One possible mode of AE/MS instrumentation would involve installing a suitable number of transducers on the surface of the structure (A) and/or in holes drilled outwards from it (B). This would require access to the structure. A second mode of instrumentation, and one which is in many cases more convenient or the only one possible, involves the installation of transducers from ground surface above the structure (C1-C4).

The type of transducer, and the location and depth of the installation, will depend on a number of factors including, the depth of the structure (d_1), the depth of the overburden (d_2), the thickness and mechanical properties of the strata overlying the structure, the expected energy of AE/MS events to be detected, the desired source location accuracy, and the funds available. It should be noted at this point that there appears to be no one single type of transducer installation which is suitable for all studies; experience has shown that the optimum system must be tailored to the specific application.

3.5.3 Transducer installation

In a number of field cases, transducers have been installed underground in the associated tunnels and mine workings. Such installations involve mechanically locking or cementing the transducers into boreholes, or clamping them to a

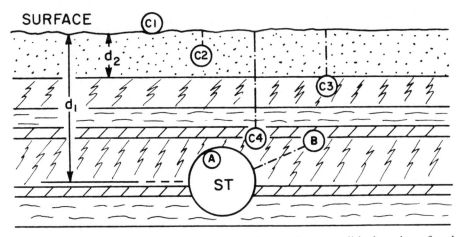

Figure 3.6. Simplified field situation illustrating various possible locations for the AE/MS transducers.

special plate previously cemented to the roof or walls of the opening. Figure 3.7 illustrates a variety of methods of underground transducer mounting. For further information in regard to this type of mounting, the reader is referred to the paper by Blake et al. (1974).

In a number of field situations it is more desirable to install transducers from the ground surface overlying the structure under study. This latter alternative has been used extensively by the author (see Hardy and Mowrey, 1977). A number of these are illustrated in Figures 3.8 to 3.11. Although these techniques were specifically developed for the study of structural stability of underground mines and underground gas storage reservoirs, the techniques should be equally applicable to a variety of other applications.

The various transducer surface installation techniques shown have evolved over the last 20 years. Initially, attempts were made to utilize the simplest and least expensive techniques; however, as a better appreciation of the limitations of various techniques were realized, optimum installation techniques were developed. A brief review of the important features of each of the installation techniques illustrated in Figures 3.8 to 3.11 follows.

1. *Near surface mounting techniques*: In general, it has been found that surface and shallow burial mounting techniques (Fig. 3.8a,b) are unsatisfactory in most applications where the structure to be monitored is deep (> 200 m). It is expected, however, that for structures less than 100 m deep, in areas

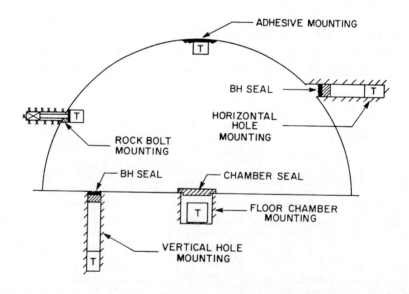

UNDERGROUND TRANSDUCER MOUNTING

Figure 3.7. Various methods employed for underground mounting of AE/MS transducers.

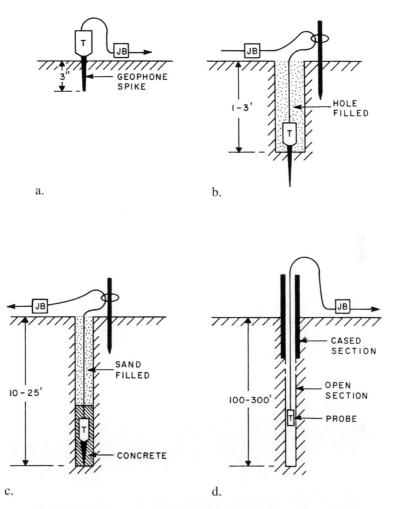

Figure 3.8. AE/MS transducer installation techniques for depths less than 100 m. (T – transducer, JB – junction box, length conversion factors: 1" = 0.025 m and 1' = 0.305 m.). a. Surface mounting. b. Shallow burial. c. Deep burial. d. Borehole probe.

with minimum overburden, these types of transducer installation may have considerable application.

2. *Deep burial techniques*: At field sites where structures are of the order of 200 m deep, and overburden is relatively shallow (≈15 m), the use of the deep burial techniques (Fig. 3.8c) has been found to be very successful (Hardy and Mowrey, 1977). This technique is optimum if the transducer can be located directly in bedrock. It should be noted, however, that in some locations, where very thick overburden is involved (≈75 m), transducer installations as deep as 10 m have not been satisfactory.

3. *Borehole probe technique*: In field situations where thick overburdens are involved and a cased borehole can be drilled to penetrate the underlying bedrock, a borehole probe or sonde (Fig. 3.8d) has been found to be very successful. A borehole probe developed at Penn State has been described in detail in a paper by Hardy et al. (1977). Figure 3.9 illustrates the manner in which the probe is installed and Figure 3.10 shows the assembled probe ready for installation. It should be noted that such an installation system offers a considerable number of advantages over other techniques, including the ability to be easily installed at various depths in both vertical and horizontal boreholes.

Single and multi-transducer sondes are available from a number of manufacturers. These employ some form of remotely activated mechanical

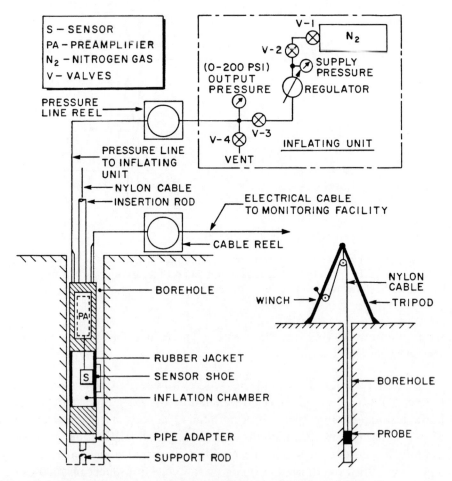

Figure 3.9. Principles of operation and installation of borehole probe.

Figure 3.10. Borehole probe ready for installation.

system for clamping the sonde at the desired depth in a cased or uncased borehole. Special sondes have been developed by researchers involved in the investigation of AE/MS activity at geothermal field sites. An earlier paper by Albright and Pearson (1982) discusses the design and use of a triaxial sonde, developed by Los Alamos National laboratory (LANL), for AE/MS studies at hot dry rock (HDR) sites. The LANL design was later modified by researchers at Tohoku University, Sendai, Japan. Figures 3.11 and 3.12 illustrate the downhole sonde system developed by Niitsuma et al. (1989). The sonde includes a number triaxial accelerometers as transducers. These

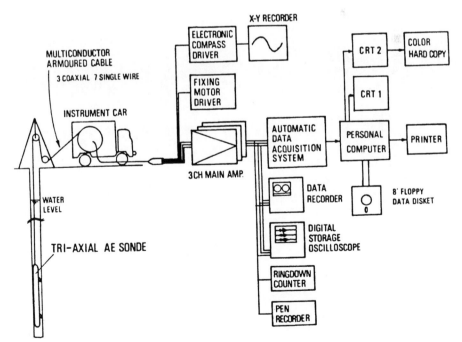

Figure 3.11. Block diagram of downhole sonde system developed for studies at geothermal field sites (after Niitsuma et al., 1989).

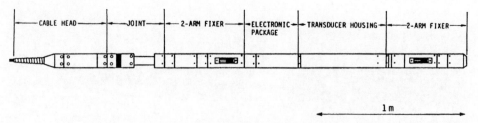

Figure 3.12. Simplified diagram of downhole sonde showing various sections and associated dimensions (after Niitsuma et al., 1989).

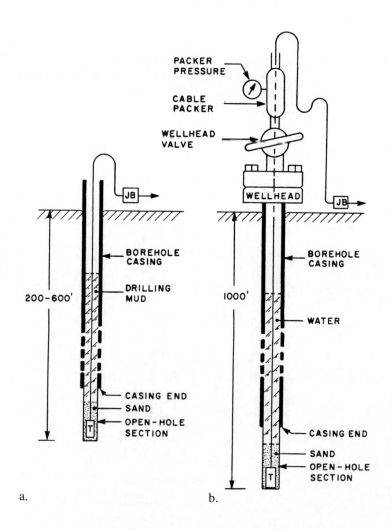

Figure 3.13. Very deep AE/MS transducer installation techniques. (T – transducer, JB – junction box, length conversion factor: 1′ = 0.305 m.). a. Deep borehole mounting. b. Pressurized borehole mounting.

and the associated preamplifiers are located in the "electronics package" noted in Figure 3.12. Two sets of 2-arm fixers are used to clamp the sonde to the borehole wall at the desired depth. The sonde was designed to operate at temperatures up to 150°C and pressures up to 20 Mpa. The attached cable allows the sonde to be installed at depths exceeding 1000 meters.

4. *Very deep techniques*: When dealing with deep structures ($\approx$300 m) or structures located in areas with thick overburden, it may be necessary to utilize a sonde system such as that illustrated in Figure 3.11 or one of the very deep techniques illustrated in Figure 3.13. These techniques, although expensive, have proven successful in situations where other techniques were found to be unsatisfactory (Hardy et al., 1981).

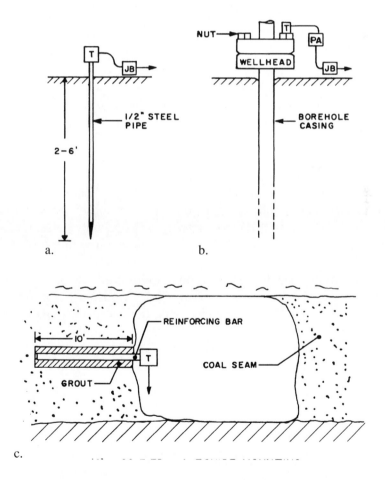

Figure 3.14. Various indirect methods for AE/MS transducer installation. (T – transducer, JB – junction box, length conversion factors: $1'' = 0.025$ m and $1' = 0.035$ m.). a. Waveguide mounting. b. Wellhead mounting. c. Modified waveguide mounting.

A series of indirect methods of mounting AE/MS transducers have also been utilized by a number of workers. In this type of installation metal rods (usually steel) are installed in the structure understudy to the depth at which AE/MS activity is to be monitored, or use is made of an in-place metal structures such as a borehole casing or reinforcing rods. Figure 3.14 illustrates a number of such indirect installation techniques.

In early Penn State studies (Hardy, 1975) attempts were made to make use of both waveguide and wellhead techniques (Fig. 3.14a,b), with little success. In contrast, Koerner et al. (1980) and McElroy (1979) have found the waveguide method suitable for use in soil structures. In recent AE/MS studies carried out in a deep West German coal mine, Hahnekamp and Heusinger (1980) have investigated the suitability of the modified waveguide installation technique illustrated in Figure 3.14c. In this mine, steel reinforcing bars, grouted into specially drilled boreholes, are routinely installed as a method of stabilizing the coal in main entries. Since boreholes in coal normally close rapidly due to high stresses, the attachment of transducers to threaded end-sections of such reinforcing bars, rather than placing them in boreholes, appears to provide both an economical and stable technique.

In most cases, however, indirect transducer installation techniques similar to those described are undesirable and should only be utilized as a last resort. In the first place, both the geometry and character of the waveguide, and the interaction between the waveguide and the surrounding soil or rock, can seriously modify any detected AE/MS signals. Secondly, unless it is possible to achieve mechanical isolation between the waveguide and the surrounding material, mechanical action (e.g. friction) between these two elements may itself generate considerable AE/MS activity. Finally, although the technique may be useful for general monitoring, the uncertainties introduced by the technique suggests that an array of transducers installed in this manner could not be utilized for meaningful location monitoring. Further consideration relative to the use of waveguides will be included later in Chapter 6.

There is no doubt that further studies to develop optimum AE/MS transducer installation techniques for specific field conditions are required. Past experience has indicated that most field sites are unique and therefore optimum array geometry and installation techniques can only be determined after carrying out a preliminary AE/MS evaluation.

3.6   FIELD MONITORING FACILITIES

3.6.1   General

After the required transducer or transducer array has been installed, suitable facilities must be provided to monitor the associated AE/MS signals. An example of such facilities are those developed by Obert and Duvall for early underground mining studies. As noted in Figure 3.15 these consisted of a "geophone" (containing a Rochell salt cantilever element enclosed in a moisture tight metal cylinder) the output of which was fed to a battery-operated amplifier and signal shaping circuits (discriminators), and then to a dual-channel film-type recorder. A set of earphones was also included to provide an audible means of monitoring AE/MS activity. In practice the geophone was located underground in a suitable borehole. This monitoring system was mainly sensitive to disturbances in the 1000 Hz frequency range due for the most part to the characteristics or the geophone itself. Using these facilities studies of the character and frequency of occurrence of AE/MS events in rock burst prone mining areas were carried out.

A review of the literature indicates that since the early studies of Obert and Duvall a wide variety of AE/MS field monitoring facilities have been developed by various workers in the geotechnical field (see for example, Hardy, 1989, 1995, 1998; Hardy and Leighton, 1977, 1980, 1984). Malott (1984) discusses the theoretical limitations of such systems. The main purpose of the following section is to provide an appreciation of the basic form of the necessary AE/MS monitoring facilities. For those interested in more detail on transducers and monitoring systems, the reader is directed to texts dealing with measurement systems (e.g., Doebelin, 1975).

In general, as illustrated in Figure 3.16a, a single channel system regardless of its apparent complexity consists of only three major sections, namely: the transducer, the signal conditioning unit, and the readout unit. In such a system, AE/MS activity is detected by the transducer, the resulting electrical signals

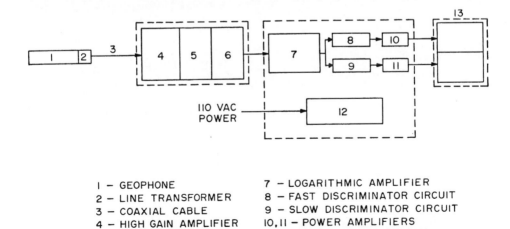

```
1 – GEOPHONE                7 – LOGARITHMIC AMPLIFIER
2 – LINE TRANSFORMER        8 – FAST DISCRIMINATOR CIRCUIT
3 – COAXIAL CABLE           9 – SLOW DISCRIMINATOR CIRCUIT
4 – HIGH GAIN AMPLIFIER     10,11 – POWER AMPLIFIERS
5 – ATTENUATOR              12 – SYSTEM POWER SUPPLY
6 – BAND PASS FILTERS       13 – DUAL CHANNEL RECORDER
```

Figure 3.15. AE/MS field monitoring facilities used by Obert and Duvall in early underground mining studies (after Obert and Duvall, 1942).

are suitably modified by the signal conditioning section, and finally displayed and/or recorded in the readout section. In general, two totally different types of signal conditioning are in common use; in this text these will be denoted as "basic" and "parametric".

As noted earlier in the text, multi-channel monitoring systems are normally required to obtain meaningful field data and a simplified four-channel system is illustrated in Figure 3.16b. It is important to note that in most cases such a system merely involves additional transducers and signal conditioning units, with the total data from all channels being recorded on a single multi-channel readout system. Regardless of the number of monitoring channels involved, the various components of the system must be selected to provide the frequency response, signal-to-noise ratio (SNR), amplification, parametric processing, and data recording capacity necessary for the specific study.

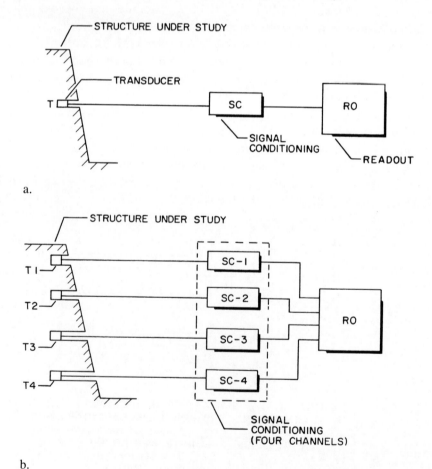

Figure 3.16. Simplified block diagrams of a single- and multi-channel AE/MS monitoring facility. a. Single channel system. b. Multi-channel system.

3.6.2 Signal conditioning

As indicated in the previous section, two different types of signal conditioning are commonly in use in AE/MS monitoring facilities, namely, basic and parametric. Figure 3.17 illustrates a simplified block diagram of a basic system. In such a system AE/MS signals detected by a suitable transducer (e.g., a geophone) are first passed through a preamplifier located near the transducer. A field cable carries the amplified signal to the monitoring facility where the signal is further amplified, filtered, and finally recorded on a multi-channel tape recorder or digital recording system. The important feature of the basic signal conditioning system is that the AE/MS data, although amplified and filtered, still retains much of its original analog form. Some workers refer to this as "full-waveform" recording.

In contrast to the basic system, the parametric system incorporates additional signal conditioning facilities which further process the amplified and filtered AE/MS signals to provide a series of specific parameters. For example, Figure 3.18 illustrates the block diagram of a single-channel monitoring system incorporating a parametric signal conditioning system which automatically provides event rate data. Such systems are available to provide one or more AE/MS parameters such as arrival time, total number of events, event rate, event energy, amplitude distribution, etc. For the most part, these parameters are determined using digital techniques. Although such signal conditioning systems are extremely useful in certain specific situations, they have a serious disadvantage in that the original analog form of the AE/MS data is permanently lost.

A number of systems utilizing parametric signal conditioning, some containing micro-processor units and capable of multi-channel monitoring, are commercially available. However, in light of the general complexity of most geotechnical field situations, it is the author's opinion that AE/MS monitoring systems incorporating basic-type signal conditioning should be employed

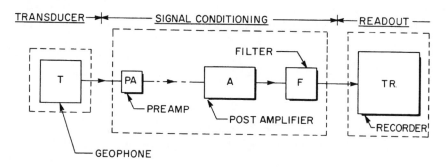

Figure 3.17. Single channel AE/MS monitoring facility incorporating a basic signal conditioning system.

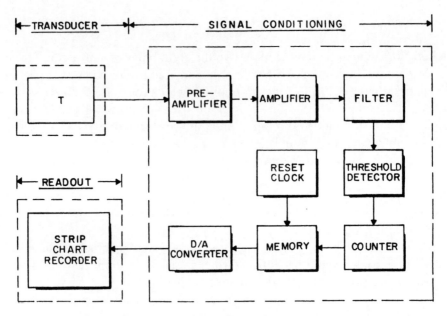

Figure 3.18. Single channel AE/MS monitoring facility incorporating a parametric signal conditioning system.

during preliminary studies at any new field site. Following a suitable study of the analog data obtained during such studies, and after the precise character of the associated AE/MS, background and cultural signals are evaluated, an intelligent decision can then be made as to the most suitable instrumentation for continuing studies at that site.

3.6.3   Readout systems

Depending on the type of signal conditioning system utilized, a variety of readout systems are available for storage and display of AE/MS data. If basic-type signal conditioning is employed, the most flexible system is one incorporating a tape recorder and an associated strip chart recorder (usually a high speed U-V recorder) for detailed visual examination of the data. Storage-type oscilloscopes, transient recorders and suitably equipped personal computers are also useful for monitoring and evaluation of selected AE/MS signals. When parametric-type signal conditioning is employed, the data (e.g., event rate) may be displayed on a simple panel meter, a digital display, a printer, or a graphics unit. In more sophisticated systems, output data from either basic or parametric signal conditioning systems may be recorded in digital form. In many respects the type of readout system employed is intimately related to the method of data processing to be utilized for later analysis of the collected data.

3.6.4 Multi-channel monitoring systems

As noted earlier, systems similar to those shown in Figures 3.17 and 3.18 are suitable for monitoring data from only a single transducer. In most field cases, however, the actual source of the AE/MS activity is required and this can only be obtained using data from an array of transducers (normally a minimum of five), and a multi-channel monitoring system is therefore required. Details on the design and construction of a variety of such systems are available in the literature (e.g., Cete, 1977; Farstad et al., 1976; Godson et al., 1980; Kimble, 1977; Leighton and Steblay, 1977; Trombic and Zuberek, 1977; Will, 1980).

The most straightforward type of multi-channel monitoring system is the basic/analog type in which the AE/MS data is recorded in an analog format. The mobile monitoring facility developed at Penn State in the early 1970s (Kimble, 1977) is a typical example of such a system. Figure 3.19 shows a block diagram of the system which was originally designed to handle up to seven channels of AE/MS data but was later expanded to 10 channels. System power was provided from a motor generator or from local power (if available), and in order to carry

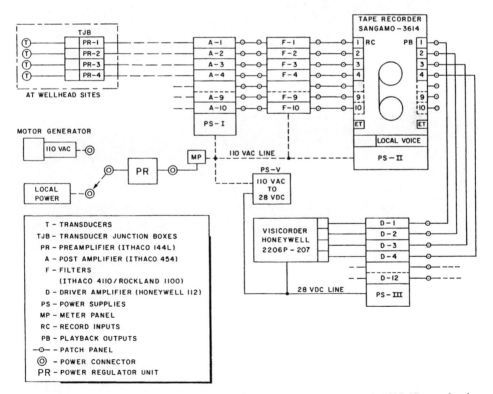

Figure 3.19. Block diagram of Penn State mobile multi-channel AE/MS monitoring facility incorporating basic/analog signal conditioning and recording (after Hardy et al., 1981).

Figure 3.20. Penn State mobile multi-channel AE/MS monitoring facility.

out measurements at a number of different locations, the completed facility was housed in a small van and associated trailer. Figure 3.20 shows a photograph of the completed monitoring facility removed from the van.

In recent years, with the advent of relatively low cost digital equipment, and the need for greater monitoring channel capacity associated with field studies of larger and more complex structures, utilization of multi-channel digital-type monitoring systems has increased. Many of these systems are either of the basic-type (e.g., Cete, 1977; Will, 1980) providing only wave form data, or of the parametric-type (e.g., Leighton and Steblay, 1977; Godson et al., 1980) providing such parameters as event amplitude and location. However, others are of a more sophisticated nature providing both basic and detailed parametric information (e.g., Batchelor et al., 1983; Brink, 1984).

A typical basic/digital monitoring system is illustrated in Figure 3.21, and a paper by Blake (1977) discusses the development of such systems for rock burst monitoring. A system of this type has been utilized by Will (1980) for monitoring the data from a set of 12 geophones installed in a German coal mine. The initial (analog) stage of the signal conditioning section is similar

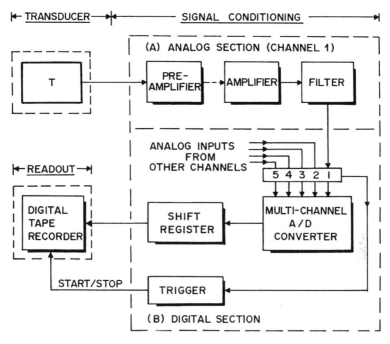

Figure 3.21. Block diagram of a typical basic/digital AE/MS monitoring facility.

to that used for typical AE/MS field studies. In this system the output from a total of 12 such channels are then multiplexed, digitized and fed, via a shift register, to a digital tape recorder. Although an additional electronics system is required for retrieval of the recorded digital data, the digital tape recorder basically represents the readout section of the monitoring system. The purpose of the shift register is to provide a time delay ($\Delta t \cong 1.8$ s) during which the data is temporally held in the shift register. The on-off status of the tape recorder is controlled by an associated trigger circuit, which is activated by the amplitude and sequence of the analog signals occurring at the input to the digital section of the signal conditioning system. If the character of these signals satisfies a preset criteria the tape recorder is turned on and the data stored in the shift register are dumped to the tape recorder. In this way only valid AE/MS events are recorded.

It is important to note that in the basic/digital monitoring system the general character of the individual AE/MS events are retained throughout the system, and these events may be recovered at a later time by suitable playback procedures. In contrast, Figure 3.22 illustrates a block diagram of a typical parametric/digital monitoring system similar to that developed by Godson et al. (1980) for monitoring AE/MS activity in an Australian, hardrock mine. Here the analog signals from up to 32 transducers are first led into a pre-processor unit which provides digital signals for each channel containing arrival-time and amplitude data only. These digital signals are then processed by an on-line computer and critical parametric data such as source location coordinates, source amplitude, etc., are both stored and outputed to an associated printer and plotter. With such monitoring systems there is no possibility of retrieving the original events themselves since they are effectively lost at the preprocessor input.

While parametric/digital monitoring systems are useful in a number of field applications there is little doubt that the most desirable AE/MS field monitoring system would incorporate the best features of both the basic and parametric digital systems. Unfortunately, such a hybrid system is both complex and expensive, particularly if a large number of data channels are involved. Figure 3.23 is a simplified block diagram for a system of this type. Such systems have been developed by a number of workers, including Brink (1984) for the detailed study of AE/MS data associated with rock bursts in deep South African gold mines, and Niitsuma et al. (1989) for the investigation of the behavior of geothermal reservoirs. In such hybrid-digital systems, analog signals from a number of data channels are first digitalized by a multi-channel A/D converter. These digital signals are then fed to an on-line computer where

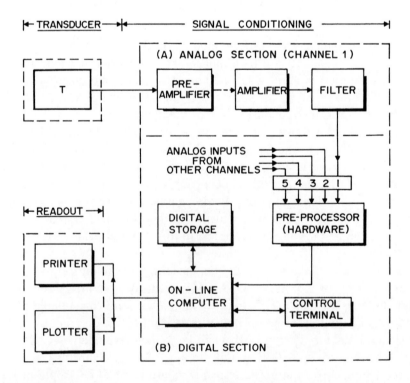

Figure 3.22. Block diagram of a typical parametric/digital AE/MS monitoring facility.

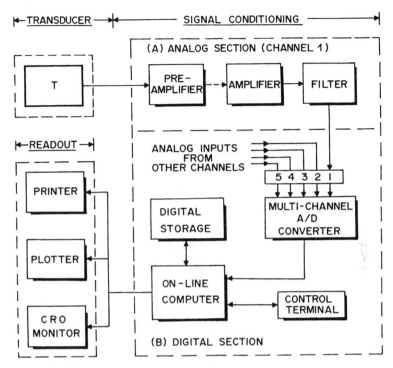

Figure 3.23. Block diagram of a typical hybrid-digital AE/MS monitoring system.

valid AE/MS signals are selected, permanently stored and displayed on a CRO monitor. During the data gathering process the system also computes a variety of parameters, such as source location coordinates, event time, source magnitude, etc., stores this information and displays it on a variety of periphals such as printers and plotters. The system thus computes and stores a wide range of parametric data as well as retaining in storage the detailed character of the original AE/MS signals.

Clearly the advent of inexpensive, high speed personal computer (PC) facilities in the last few years are changing the way that AE/MS data is being collected, stored and analyzed. A number of manufacturers, including R. C. Electronics Inc., California, and Gage Applied Sciences, Inc., Montreal, Canada, supply multi-channel A/D and signal processing cards for PC application along with associated software. These so-called "computerscope"/ "compuscope" facilities are very applicable to AE/MS monitoring and will be discussed further under laboratory monitoring systems in Chapter 4.

Other more specialized computer-based systems have been under development, including those by Majer et al. (1984), Will (1984) and Yamaguchi et al. (1985). For example a recent paper by Mendecki (1993) describes the concept of a quantitative, real-time AE/MS field monitoring system, denoted by the writer as an Integrated Seismic System (ISS). In this system a number of "intelligent" transducers are connected to a central computer. The system is capable of automatic generation of P- and S-wave arrival time data, source location, and spectral parameters, as well as detailed moment tensor analysis, an increasingly popular analytical technique.

REFERENCES

Albright, J.N. and Pearson, C.F. 1982. Acoustic Emissions as a Tool for Hydraulic Fracture Location: Experience at the Fenton Hill Hot Dry Rock Site, *Society Petroleum Engineering Journal*, August 1982, pp. 523-530.

Anon. 1984. Geophone Notes, Personal Communication, Sercel Inc. (originally Mark Products, U.S., Inc.)., Houston, Texas, March 1984.

Batchelor, A.S., Baria, R. and Hearn, K. 1983. Monitoring the Effects of Hydraulic Stimulation by Microseismic Event Location: A Case Study, Preprint No. 12109, *Society of Petroleum Engineers, 58th Annual Technical Conference*, San Francisco, October 1983.

Blake, W. 1977. Design, Installation and Operation of Computer Controlled Rock Burst Monitoring Systems, *Proceedings First Conference on Acoustic Emission/ Microseismic Activity in Geologic Structures and Materials*, The Pennsylvania State University, June 1975, Trans Tech Publications, Clausthal-Zellerfeld, Germany, pp. 157-167.

Blake, W., Leighton, F. and Duvall, W.I. 1974. *Microseismic Techniques for Monitoring the Behavior of Rock Structures*, Bulletin 665, U.S. Bureau of Mines.

Brink, A.V.Z. and O'Conner, D. 1984. Rock Burst Prediction Research–Development of a Practical Early-Warning System, *Proceedings Third Conference on Acoustic Emission/Microseismic Activity in Geologic Structures and Materials*, The Pennsylvania State University, October 1981, Trans Tech Publications, Clausthal-Zellerfeld, Germany, pp. 269-282.

Cete, A. 1977. Seismic Source Location in the Ruhr District, *Proceedings First Conference on Acoustic Emission/Microseismic Activity in Geologic Structure and Materials*, The Pennsylvania State University, June 1975, Trans Tech Publications, Clausthal-Zellerfeld, Germany, pp. 231-241.

Doebelin, E.O. 1975. *Measurement Systems*, McGraw-Hill Book Company, New York.

Dowding, C.H. 1985. *Blast Vibration Monitoring and Control*, Prentice-Hall Inc., Englewood Cliffs, N.J., pp. 197-226.

Dresen, L. and Ruter, H. 1995. *Seismic Coal Exploration, Part B: In-Seam Seismics*, Pergamon Press, Elsevier Science Inc., New York, pp. 279-323.

Farstad, A.J., Kalvels, D., Kehrman, R.F., Fisher, C. Jr. and Leschek, W.C. 1976. *Microseismic Roof Fall Warning System*, NTIS Publication No. PB -262-868, U.S. Department of Commerce.

Galperin, E.I., Nersesov, I.L. and Galperina, R.M. 1986. *Borehole Seismology*, Reidel Publishing Co., Boston, pp. 5-20.

Godson, R.A., Bridges, M.C. and McKavanagh, B.M. 1980. A 32-Channel Rock Noise Source Location System, *Proceedings Second Conference on Acoustic Emission/ Microseismic Activity in Geologic Structures and Materials*, The Pennsylvania State University, November 1978, Trans Tech publications, Clausthal-Zellerfeld, Germany, pp. 117-161.

Hahnekamp, H.G. and Heusinger, P.P. 1980. Personal Communication, Forschungs-instit des Steinkolnbergbauvereins, Bergbau-Forschung GmbH, Essen, West Germany, June 1980.

Hardy, H.R. Jr. 1975. *Feasibility of Utilizing Microseismic Techniques for the Evaluation of Underground Gas Storage Reservoir Stability*, AGA Catalog No. L19725, American Gas Association, Inc., Arlington, Virginia.

Hardy, H.R., Jr. (ed.) 1989. *Proceedings, Fourth Conference on Acoustic Emission/ Microseismic Activity in Geologic Structures and Materials*, The Pennsylvania State University, October 1985, Trans Tech Publications, Clausthal-Zellerfeld, Germany.

Hardy, H.R., Jr. (ed.) 1995. *Proceedings, Fifth Conference on Acoustic Emission/ Microseismic Activity in Geologic Structures and Materials*, The Pennsylvania State University, June 1991, Trans Tech Publications, Clausthal-Zellerfeld, Germany.

Hardy, H.R. Jr. (ed.) 1998. *Proceedings, Sixth Conference on Acoustic Emission/ Microseismic Activity in Geologic Structures and Materials*, The Pennsylvania State University, June 1996, Trans Tech Publications, Clausthal-Zellerfeld, Germany.

Hardy, H.R. Jr. and Leighton, F.W. (eds.) 1977. *Proceedings First Conference on Acoustic Emission/Microseismic Activity in Geologic Structures and Materials*, The Pennsylvania State University, June 1975, Trans Tech Publications, Clausthal-Zellerfeld, Germany.

Hardy, H.R. Jr. and Leighton, F.W. (eds.) 1980. *Proceedings Second Conference on Acoustic Emission/Microseismic Activity in Geologic Structures and Materials*, The Pennsylvania State University, November 1978, Trans Tech Publications, Clausthal-Zellerfeld, Germany.

Hardy, H.R. Jr. and Leighton, F.W. (eds.) 1984. *Proceedings Third Conference on Acoustic Emission/Microseismic Activity in Geologic Structures and Materials*, The Pennsylvania State University, October 1981, Trans Tech Publications, Clausthal-Zellerfeld, Germany.

Hardy, H.R. Jr. and Mowrey, G.L. 1977. Study of Underground Structural Stability Using Near Surface and Down Hole Microseismic Techniques, *Proceedings Symposium on Field Measurements in Rock Mechanics*, Zurich, 1977, A. A. Balkema Co., Rotterdam, Vol. I, pp. 75-92.

Hardy, H.R. Jr., Comeau, J.W. and Kim, R.Y. 1977. Development of a Microseismic Borehole Probe for Monitoring the Stability of Geologic Structures, *Proceedings 18th U.S. Symposium on Rock Mechanics*, Keystone 1977, F.D. Wang and G.B. Clark (eds.), Colorado School of Mines Press, pp. 3A5-1 to 3A5-7.

Hardy, H.R. Jr., Mowrey, G.L. and Kimble, E.J. Jr. 1981. *A Microseismic Study of an Underground Natural Gas Storage Reservoir, Volume I – Instrumentation and Data Analysis Techniques, and Field Site Details*, AGA Catalog No. L51396,American Gas Association, Arlington, VA, 343 pp.

Kimble, E.J. Jr. 1977. Development of Mobile Microseismic Field Equipment, *Proceedings First Conference on Acoustic Emission/Microseismic Activity in Geologic Structures and Materials*, The Pennsylvania State University, June 1975, Trans Tech Publications, Clausthal-Zellerfeld, Germany, pp. 339-356.

Koerner, R.M., Lord, A.E. Jr. and McCabe, W.M. 1980. The Challenge of Field Monitoring of Soil Structures Using AE Methods, *Proceedings Second Conference on Acoustic Emission/Microseismic Activity in Geologic Structures and Materials*, The Pennsylvania State University, November 1978, Trans Tech Publications, Clausthal-Zellerfeld, Germany-Zellerfeld, pp. 275-289.

Leighton, F. and Steblay, B.J. 1977. Applications on Microseismics in Coal Mines, *Proceedings First Conference on Acoustic Emission/Microseismic Activity in*

Geologic Structures and Materials, The Pennsylvania State University, June 1975, Trans Tech Publications, Clausthal-Zellerfeld, Germany, pp. 205-229.

Majer, E.L., King, M.S. and McEvilly, T.V. 1984. The Application of Modern Seismological Methods to Acoustic Emission Studies in a Rock Mass Subjected to Heating, *Proceedings Third Conference on Acoustic Emission/Microseismic Activity in Geologic Structures and Materials*, The Pennsylvania State University, October 1981, Trans Tech Publications, Clausthal-Zellerfeld, Germany, pp. 499-516.

Malott, C. 1984. Theoretical Limitations of Microseismic Transducer Systems, *Proceedings Third Conference on Acoustic Emission/Microseismic Activity in Geologic Structures and Materials*, The Pennsylvania State University, October 1971, Trans Tech Publications, Clausthal-Zellerfeld, Germany, pp. 681-693.

McElroy, J.W. 1979. Acoustic Emission Inspection of Buried Pipelines, *Acoustic Emission Monitoring of Pressurized Systems*, STP 697, American Society for Testing and Materials, Philadelphia, Pennsylvania, pp. 47-59.

Meister, D. 1980. Microacoustic Studies in Salt Rock, *Proceedings Second Conference on Acoustic Emission/Microseismic Activity in Geologic Structures and Materials*, The Pennsylvania State University, November 1978, Trans Tech Publications, Clausthal-Zellerfeld, Germany, pp. 259-273.

Mendecki, A.J. 1993. Real Time Quantitative Seismology in Mines, *Rock Bursts and Seismicity in Mines 93*, R.P. Young (ed.), A.A. Balkema, Rotterdam, pp. 287-295.

Nittsuma, H., Nakatsuka, K., Chubachi, N., Yokoyama, H. and Takanohashi, M. 1989. Downhole AE Measurement of a Geothermal Reservoir and its Application to Reservoir Control, *Proceedings Fourth conference on Acoustic Emission/Microseismic Activity in Geologic Structures and Materials*, The Pennsylvania State University, October 1985, Trans Tech Publications, Clausthal, Germany, pp. 475-489.

Obert, L. and Duvall, W. 1942. *Use of Subaudible Noises for the Prediction of Rock Bursts, Part II*, RI 3654, USBM.

Oh, E. 1998. A Computer-Based Calibration System for Low Frequency Acoustic Emission Transducers, *Proceedings Sixth Conference on Acoustic Emission/Microseismic Activity in Geologic Structures and Materials*, The Pennsylvania State University, Trans Tech Publications, Clausthal-Zellerfeld, Germany, pp. 471-488.

Power, D.V. 1977. Acoustic Emissions Following Hydraulic Fracturing in a Gas Well, *Proceedings First Conference on Acoustic Emission/Microseismic Activity in Geoloigic Structures and Materials*, The Pennsylvania State University, June 1975, Trans Tech Publications, Clausthal-Zellerfeld, Germany, pp. 291-308.

Schuck, L.Z. and Keech, T.W. 1977. Monitoring Acoustic Emission from Propagating Fractures in Petroleum Reservoir Rocks, *Proceedings First Conference on Acoustic Emission/Microseismic Activity in Geologic Structures and Materials*, The Pennsylvania State University, June 1975, Trans Tech Publications, Clausthal-Zellerfeld, Germany, pp. 309-338.

Schulte, L. 1980. *Status of Acoustic Detection by Microseismic Monitoring*, Solution Mining Research Institute, Annual Fall Meeting, Minneapolis, October 1980.

Trombik, M. and Zuberek, W. 1977. Microseismic Research in Polish Coal Mines, *Proceedings First Conference on Acoustic Emission/Microseismic Activity in Geologic Structures and Materials*, The Pennsylvania State University, June 1975, Trans Tech Publications, Clausthal-Zellerfeld, Germany, pp. 170-194.

Yamaguchi, K., Hamada, T., Ichikawa, H. and Oyaizu, H. 1985. Advanced Acoustic Emission Monitoring System by Distributed Processing of Waveform Microdata and the System Configuration, Abstracts Second International Conference on Acoustic

Emission, Lake Tahoe, October/November 1985, Special Supplement *Journal Acoustic Emission, Vol. 4*, No. 2/3, pp. S325-S328.

Will, M. 1980. Seismoacoustic Activity and Mining Operations, *Proceedings Second Conference on Acoustic Emission/Microseismic Activity in Geologic Structures and Materials*, The Pennsylvania State University, November 1978, Trans Tech Publications, Clausthal-Zellerfeld, Germany, pp. 191-209.

CHAPTER 4

Laboratory monitoring systems

4.1 INTRODUCTION

In the previous chapter the general concepts utilized in AE/MS field monitoring
systems have been discussed in considerable detail. Many of these same
concepts are also applicable to laboratory monitoring systems and this chapter
will therefore concentrate mainly on those additional concepts which apply
specifically to such systems. A paper by Mobley et al. (1987) provides a useful
review of facilities suitable for AE/MS laboratory studies through the late
1980s.

   AE/MS laboratory studies involve the detection and processing of events
occurring in a finite body (specimen or model), in contrast to field studies
where infinite or semi-infinite bodies are generally involved. It should be noted,
however, that certain types of small scale studies carried out in the field, for
example, in-situ mechanical property tests, detailed studies of small sections
of underground structures, etc., may also be classed as laboratory studies, at
least in respect to their dimensional scale. As a result, in each field studies,
laboratory-type rather than field-type monitoring procedures may be more
applicable.

4.2 GENERAL CONSIDERATIONS

Although the methods of detection and processing of AE/MS signals in the
laboratory are similar to those used in the field it should be appreciated that
important differences exist due to a number of factors, including smaller source
dimensions, the finite nature of the structure under study and shorter transducer-
to-source distances. In comparison to AE/MS field data, data detected during
laboratory studies will generally exhibit the following characteristics:
1. high dominant signal frequencies,
2. low signal amplitudes,
3. high event rates, and

4. signal complexities due to stress wave reflections from specimen or model
   boundaries.

These factors have important implications in regard to the design of suitable
monitoring systems for laboratory use.

Since laboratory structures (specimens or models) will have a finite nature
(e.g., a mean dimension in the range 0.02-2.0 m), the dimensions and the
available energy of the major AE/MS sources (e.g., inter- and intra-grain
failure, crack initiation and propagation, etc.) will be small. This will result in
AE/MS signals of high dominant frequency and low amplitude. However, since
the transducer-to-source distances are small, attenuation will generally be low
(compared to that associated with field studies) and, due to the fact that higher
signal-to-noise ratios are generally possible in the laboratory, a large number
of events will be detectable.

In general, laboratory structures will consist of a single type of material
eliminating wave propagation complexities due to reflection and refraction at
boundaries within the structure itself. However, the finite nature of the structure
will introduce additional problems due to reflections at free boundaries and
at interfaces between the test structure and the associated test fixtures (e.g.,
loading platens).

Finally, due to the finite nature of the structure, care must be taken in respect
to the intrinsic resonant frequencies of the structure and the associated test
fixtures, and also the effect of the transducer (weight and location) on these
frequencies. Although in tests on geologic materials this aspect has not been
investigated in detail, a number of studies have indicated that the frequency
spectra of AE/MS events detected in laboratory structures do reflect these
intrinsic resonant frequencies, at least to a limited degree.

4.3   LABORATORY TRANSDUCERS

4.3.1   Transducer types

As noted earlier, most laboratory studies in the geotechnical area involve the
detection of AE/MS events with dominant frequencies in the range 50 kHz
- 500 kHz, although studies at both lower and higher frequencies have been
carried out. Based on the availability of commercial transducers and associated
monitoring systems, however, the range 100 kHz-300 kHz is probably the most
common. Considering the general discussion of AE/MS transducers presented
earlier in Chapter 2, it would seem apparent that the most suitable type of
transducer for laboratory use would be some form of conventional piezoelectric
accelerometer. Unfortunately relatively few commercial accelerometers are
suitable for use at such high frequencies and those available have extremely

low sensitivities. It should be noted, however, that conventional accelerometers are well suited for laboratory studies below 50 kHz, particularly where flat frequency response is required for spectral analysis. At higher frequencies two general types of transducers are normally employed, namely: a piezoelectric element permanently mounted in a suitable housing and denoted here as an "AE-transducer", and individual piezoelectric elements attached directly to the structure under study. Limited use has also been made of semi-conductor strain gages; however, their sensitivity is considerably lower than that of the other type of transducers. In the following sections each of the four types of transducers will be briefly discussed. An interesting text on electrical transducers in general is that by Norton (1969).

Accelerometers: Many of the early, laboratory studies (prior to 1979) involved studies at relatively low frequencies (f < 20 kHz) and utilized conventional accelerometers as AE/MS transducers. Figure 4.1 illustrates the general form of a typical accelerometer. The active element consists of a number of piezoelectric discs on which rests a relatively heavy mass. The mass is preloaded by a stiff spring and the whole assembly seated in a metal housing with a thick base. When the accelerometer is subjected to vibration the mass exerts a variable force on the discs which, due to the piezoelectric effect, develop a voltage proportional to the force and therefore to the acceleration of the mass. For frequencies much lower than the resonant frequency (f_n) of the assembly, the voltage generated is directly proportional to the acceleration to which the transducer itself is subjected.

Figure 4.2 illustrates the typical frequency response of a number of accelerometers manufactured by Bruel & Kjaer Instruments, Inc. A list of other manufacturers is presented in the appendices. It is apparent from this figure that only a few transducers are available for use above 100 kHz (even in the resonant mode). Furthermore a review of the specifications for these transducers and those available from other manufacturers indicate that the basic transducer sensitivity drops rapidly with an increase in the maximum operating frequency. For example, the ratio of the basic sensitivities for the

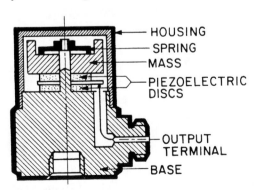

Figure 4.1. Construction of a typical accelerometer (after Anon., 1981).

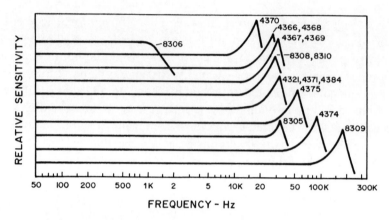

Figure 4.2. Typical frequency response curves for a number of Bruel & Kjaer accelerometers (after Anon., 1981).

Bruel & Kjaer models 4370 (f_n = 18 kHz) and 8309 (f_n – 180 kHz) is found to be approximately 250. Furthermore if a flat frequency response is required the upper operating frequency is normally limited to 1.3 f_n. As a result the two specified transducers would be limited to a useable range of approximately 0-5 kHz and 0-50 kHz, respectively.

In review then it is clear that for low frequency AE/MS laboratory studies, conventional accelerometers provide excellent results, and these are particularly useful when flat frequency response is required. In the past such transducers have been utilized in a number of low frequency AE/MS laboratory studies (e.g., Chugh, 1968; Kim, 1971). Such accelerometers, however, are not suitable for the higher frequencies normally encountered.

AE-transducers: In the last 25 years a special type of transducer, known as an "acoustic emission (AE) transducer" has become commercially available. In essence, as illustrated in Figure 4.3, the simplest form of AE-transducer consists of a single piezoelectric disc mounted on a flat sole or wear-plate, backed by a suitable damping material, and contained within a protective housing. In use, the transducer is attached to the specimen or structure under study and an AE/MS stress wave striking the wear-plate induces a stress (pressure) within the piezoelectric disc which generates an equivalent output voltage. Since the piezoelectric element in such a transducer is usually only lightly damped, it has a very high sensitivity at resonance. Most AE-transducers are utilized in the so-called resonance mode in order to achieve a high sensitivity, although a number of transducers with "flat frequency response" over a somewhat extended frequency range are available.

A number of manufacturers supply transducers in both a single-ended and differential form. The latter, which involves a somewhat more complex piezoelectric element configuration, has the advantage that it may be utilized with a suitable differential preamplifier to cancel out electrical noise and thus

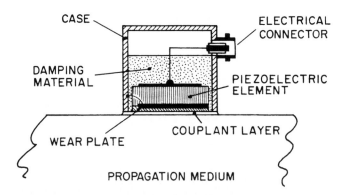

Figure 4.3. Basic construction of a simple AE-transducer.

increase the effective signal-to-noise ratio (SNR) of the overall monitoring system. AE-transducers developed by a number of companies are presently in use. A list of current manufacturers are presented in the appendix. Figure 4.4 illustrates a variety of AE-transducers designed for use in the 100 kHz – 300 kHz range. These transducers also exhibit a range of electrical connections, namely, from left to right: miniature (microdot) connector, BNC connector, and integral cable.

Figure 4.4b illustrates a commonly used Dunegan model D-140B AE-transducer of the resonant type. The prefix "D" indicates that the transducer is of the differential form. Figure 4.5 illustrates typical calibration data for the D-140B transducer obtained using the "spark impulse" calibration technique".*
In contrast, Figure 4.6 illustrates the calibration curve for a flat frequency response AE-transducer, namely, the Dunegan model D-9201.

As noted in Figures 4.5 and 4.6 the sensitivity of AE transducers are normally quoted in terms of acoustic parameters (see Norton, 1969), i.e., dB re: 1v/μbar, where 1 bar = 14.5 psi = 10^5 pascal. These sensitivity values can be converted to those values commonly utilized in the geotechnical area using the following relationship:

$$S' = 20\log_{10}S \qquad \text{(Eq. 4.1)}$$

Where S' and S are the transducer sensitivity in dB re: lv/ bar and volts/μbar, respectively. For example, the peak sensitivity of a Dunegan model D-140B transducer is listed as $S' = -75$ dB re: lv/μbar. Using Equation 4.1 it is found that $S = 177.83$ μv/μbar or 12.25 v/psi.

---

* Two types of AE-transducer calibration are presently in use, namely: spark impulse and ultrasonic calibration. Somewhat different results are obtained by the two techniques, and the general problem of AE-transducer calibration is presently under study by a number of workers.

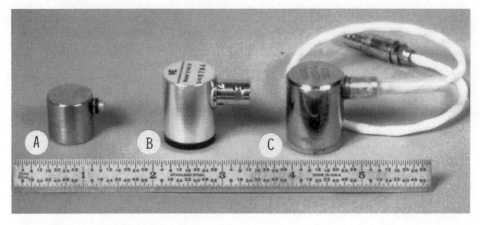

Figure 4.4. A variety of AE-transducers for operation in the range 100 kHz – 300 kHz. (Manufacturers: A – Physical Acoustics, B – Dunegan and C – Acoustic Emission Technology).

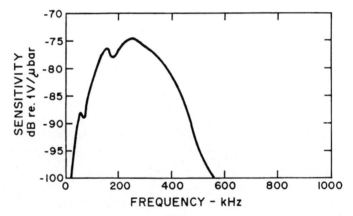

Figure 4.5. Calibration curve for Dunegan/Endevco model D-140B resonant-type AE-transducer (spark impulse calibration) (after Anon., 1983).

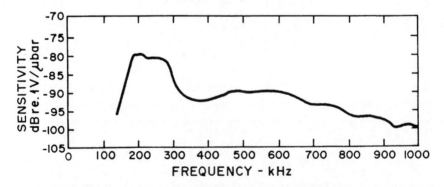

Figure 4.6. Calibration curve for Dunegan/Endevco model D-9201 flat frequency response-type AE-transducer (spark impulse calibration) (after Anon., 1983).

A wide range of AE-transducers are available from a number of manufacturers (see appendices). These include resonant- and flat frequency response-types covering a range of frequencies up to 1.0 MHz and higher. Special models are also available with built-in preamplifiers along with those for use under hostile environments, e.g., high temperature, high radiation fields, etc. Further discussion of commercially available AE-transducers are available from the manufacturers and in a number of published reports (e.g. Eitzen and Breckenridge, 1987; Mobley et al., 1987). Over the last few years, H. L. Dunegan has published a number of very useful short papers on AE-transducer design and calibration in the DECI Report, a continuing technical publication of Dunegan Engineering Consultants, Inc. Copies of these papers are available from DECI.

In recent years there has been research underway to develop a high fidelity AE-transducer that could be used as a "standard". This work was mainly carried out at the National Institute of Standards and Technology (NIST) in Gaithersburg, Maryland. The transducer utilizes a cone-shaped piezoelastic element which provides a very small aperture (1.0 – 1.5 mm), and responds to only a single variable, displacement. Papers by Proctor (1982a,b, 1986) describe the development and calibration of the high fidelity transducer. Glaser has made considerable use of this type of transducer to study fracture in rock and concrete (Glaser and Nelson 1992a,b). A recent paper by Weiss and Glaser (1998) describes the development of an embeddible sensor for monitoring hydrofrac body waves based on the NIST transducer.

Piezoelectric elements: As indicated earlier in this section stress waves associated with AE/MS activity may be detected by transducers such as accelerometers or AE-transducers attached to the structure under study. In both devices the basic sensing unit is some form of piezoelectric element. A variety of natural and artificial piezoelectric materials are available; however, barium titanate or lead-zirconate titanate (PZT) are commonly used. Under certain circumstances (e.g., space restrictions) it is more convenient to directly attach the piezoelectric element itself to the test structure.

Piezoelectric elements in the form of discs and bars, having conductive upper and lower surfaces to provide electrical contact with the piezoelectric material, are commercially available from a number of manufacturers (see appendices). The operating characteristics (e.g., resonant frequency) of such elements depend on their size and geometry. For example, disc-shaped elements, as shown in Figure 4.7a, were used by Siskind (1970) to monitor AE/MS activity in small laboratory models tested under triaxial stress. These elements, with a diameter of 0.003 m and a thickness of 0.001 m, had a resonant frequency of 1.56 MHz. Such elements are available for sensing either P- or S-waves. In general, as illustrated in Figure 4.7b, piezoelectric elements are attached to the test structure using some form of conductive epoxy. Further details in this regard will be presented later in Section 4.3.2.

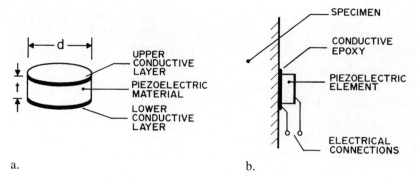

Figure 4.7. Details in regard to construction of a typical disc-type piezoelectric element and method of installation. a. Typical disc-type piezoelectric element. b. Method of installation.

Semiconductor strain gages: The use of wire- or foil-type SR-4 strain gages is common in the geotechnical area (e.g., Doebelin, 1975; Dove and Adams, 1964). These are small resistive devices which, when cemented to laboratory or field structures, allow the measurement of surface strains. The sensitivity of such devices is indicated by their so-called gage factor, and the relationship between observed changes in gage resistance and surface strain is as follows:

$$\Delta R = (RG)e \qquad\qquad (Eq.\ 4.2)$$

where ΔR is the change in gage resistance due to a surface strain e, and R and G are the nominal gage resistance and the gage factor, respectively. Standard types of wire and foil gages have gage factors in the range of 1.5 – 3.5; however, values of the order of 2.0 are most common. In use the change in strain gage resistance (ΔR) is monitored using a suitable Wheatstone bridge circuit to provide a bridge output voltage (E_o) proportional to the level of surface strain.

Dynamic surface strains occurring due to an incident stress wave could certainly be measured using conventional SR-4 strain gages provided that the stress wave, and the resulting surface strains, were of sufficient amplitude. Using suitable high gain monitoring systems, such measurements are practical for high amplitude stress waves associated with sources such as mechanical impact and explosions, where the resulting surface strains are greater than 10^{-6}. They are impractical, however, for the low amplitude stress waves associated with typical AE/MS sources, where the associated surface strains are normally of the order of $10^{-8} – 10^{-10}$ or less. In general the use of systems utilizing conventional SR-4 gages are limited by the sensitivity (i.e., gage factor) of the strain sensing element itself, rather than by the associated monitoring system.

Studies carried out by Kimble (1984) have indicated that special high sensitivity semiconductor strain gages, with gage factors greater than 100, may be successfully utilized to monitor AE/MS activity in laboratory structures.

Studies to date have utilized BLH* type SPB2-12-12 gages having a nominal gage factor of 115, a gage length of 0.003 m and a nominal resistance of 120 ohms. In the manufacture of these gages a bar-shaped element is cut from a single silicon crystal, appropriate leads are attached, and this configuration is then mounted to a Bakelite backing to improve the mechanical integrity of the gage. Tests indicate that such gages exhibit an essentially flat frequency response up to 1 MHz and higher. Figure 4.8 illustrates one of these gages as received from the manufacturer.

For use as an AE/MS transducer the gage was mounted on a brass shoe, contoured to fit to the test specimen surface, using Eastmen 910 cement. This arrangement allowed the mounted transducer to be temporarily attached to the structure under study and to be removed and reused in later tests. Figure 4.9 illustrates two such assembled transducers.

One of the major problems in using such transducers is the associated electronics necessary to convert the small changes in gage resistance to a suitable voltage for input to a conventional AE/MS monitoring system. Kimble (1984) describes the development of the necessary electronics, in particular special low noise amplifiers, illustrated in block diagram form in Figure 4.10. Using this system Kimble indicates that, on the basis of a SNR = 2, surface strains of the order of 4×10^{-8} could be detected.

AE/MS transducers incorporating semiconductor strain gages appear to offer a number of important advantages for laboratory studies in the geotechnical area, namely:
1. Flat frequency response over a wide range.
2. Directional capability, since the transducer monitors surface strains parallel to the axis of the strain gage element.

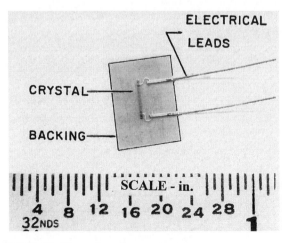

Figure 4.8. Type SPB2-12-12 semiconductor strain gage.

* BHL Electronics Inc., Waltham, Massachusetts.

3. Relatively small size (approximately 0.013 m x 0.006 m x 0.003 m thick) allowing several transducers to be located on a small specimen or structure, and making it possible to directly instrument jacketed test specimens during triaxial tests.

The transducer does have one major disadvantage in that its sensitivity is considerably less than conventional AE-transducers or piezoelectric elements.

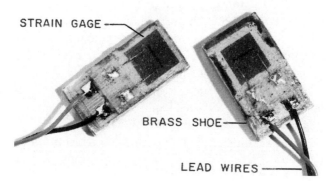

Figure 4.9. Two AE/MS transducers utilizing semiconductor strain gages.

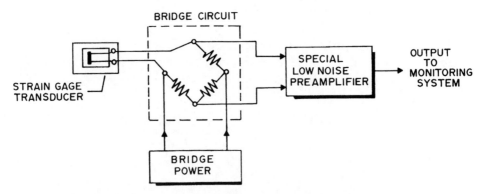

Figure 4.10. Block diagram of electronic circuit utilized with AE/MS transducers incorporating semiconductor strain gages.

4.3.2 Transducer installation

In the laboratory the various AE/MS transducers described in the previous section may be installed in a variety of ways depending on the type of transducer and the environment of the test specimen or structure. Figure 4.11 illustrates a number of such installation techniques. Further details of these will be presented in the following sections, with the techniques being subdivided in terms of their use in uniaxial and triaxial tests, and tests under hostile environments.

Uniaxial tests: Uniaxial tests under room temperature conditions are probably the most common type carried out in the rock mechanics laboratory. From an AE/MS instrumentation point-of-view such tests are the least difficult, and Figure 4.11a,b,c,f, and g illustrate a number of the associated transducer installation techniques. The two methods shown in Figure 4.11a and b are commonly utilized for both conventional accelerometers and AE-transducers. For example, studies by Chugh (1968) and Kim (1971) utilized accelerometers bonded directly to the test specimens using Eastmen 910 cement, as illustrated in Figure 4.11a. This mode of installation is convenient and allows the transducer to be easily removed from the specimen following the test for later reuse. The only disadvantage of the method appears to be the fact that during testing the deformation of the specimen may induce mechanical instabilities

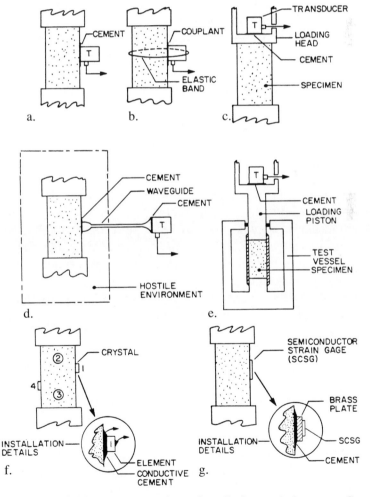

Figure 4.11. Typical laboratory transducer installation techniques. a. Cemented. b. Couplant. c. Loading head. d. Waveguide. e. Triaxial vessel. f. Piezoelectric element. g. Semiconductor strain gage.

(e.g., bond failure, microcracks, etc.) in the cement which may be detected as AE/MS activity. This potential problem may be overcome by the use of a coupling agent (a special high viscosity fluid with low acoustic attenuation characteristics) and a suitable clamping arrangement (e.g., elastic bands) to hold the transducer in place. In the past this technique, as shown in Figure 4.11b, has been used by a large number of workers (e.g., Richardson, 1978 and Roberts, 1981). It is still commonly in use.

In some situations it is convenient to mount accelerometers or AE-transducers in a separate loading head, as illustrated in Figure 4.11c. The arrangement is particularly convenient when a large number of specimens are to be tested. This technique also provides a high degree of mechanical protection for the transducer, which is an important factor if the specimens are to be tested to failure. The technique, however, has a number of serious disadvantages. First, since the transducer is not mounted directly to the specimen, serious acoustic losses may occur at the specimen-loading head interface. Furthermore, in order to keep such losses to a minimum it is not feasible to use any soft material between the specimen and the loading head as a means of providing uniform loading conditions. Secondly, the contact region between the specimen and the loading head is normally a region of high AE/MS activity due to the presence of mechanical instabilities associated with specimen misalignment, end-surface irregularities, and frictional effects due elastic mismatch between the specimen and the loading head. Since the transducer is located very close to this unstable region even low-level AE/MS events occurring here may overshadow more meaningful events occurring in the body of the specimen.

As illustrated in Figure 4.11f and g, piezoelectric elements or semiconductor strain gage transducers may also be utilized in uniaxial tests. The former have been employed extensively in AE/MS studies on geologic materials and offer the advantage of permitting a number of transducers to be attached to relatively small specimens. For example, during studies by Rothman (1975), an array of seven such PZT elements were mounted on a specimen approximately 0.029 m in diameter and 0.064 m in length, to provide source location data during short-term creep tests on sandstone specimens. In his studies, PZT elements (rectangular bars 0.0064 m long, 0.0032 m wide, and 0.0013 m thick) were first electrically wired and potted in epoxy resin, and then bonded to the specimen using Eastmen 910 cement. Since the specimens were tested to failure the preliminary epoxy potting procedure was necessary to protect the PZT elements so that they could be reused in later tests. A detailed description of the procedures used for protecting and installing the PZT elements are described elsewhere (Rothman, 1975). In other studies PZT elements have been bonded directly to the test specimens using a conductive epoxy which also provides one of the electrical connections to the piezoelectric element.

The method of installation of AE/MS transducer incorporating semiconductor strain gages is illustrated in Figure 4.11g. Normally, Eastmen 910 cement has been utilized, and experience had indicated that such transducers may be easily removed and reused in later tests.

Triaxial tests: The application of AE/MS transducers for tests under triaxial conditions (using fluid confinement) presents a number of special problems, since the specimen is normally surrounded by a rubber or plastic jacket and located within a steel test vessel. Such tests normally preclude the direct application of accelerometers or AE-transducers. However, as illustrated earlier in Figure 5.11e, indirect mounting of such transducers, in either an associated loading head or on the specimen loading piston itself, may be employed. In some studies (e.g., Siskind, 1970) a PZT element, bonded directly to the interior, bottom surface of the loading head, has also been utilized.

Indirect mounting techniques, however, introduce similar difficulties to those discussed earlier in the case of uniaxial tests where the arrangement illustrated in Figure 4.11c is utilized. In the triaxial case additional complications arise due to the greater distance between the transducer and the test specimen, and the more complex form of the loading arrangement. As a result, a series of secondary AE/MS events are generated due to reflection of the primary event at various boundaries within the loading system.

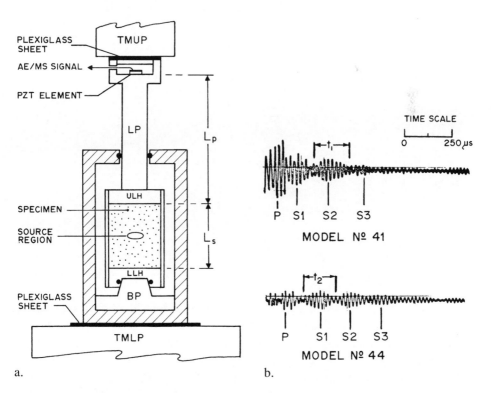

Figure 4.12. Simplified triaxial test facility and typical observed AE/MS signals associated with tests on gas storage reservoir models (after Siskind, 1970). a. Simplified triaxial test facility. b. Typical AE/MS signals.

To illustrate this problem in more detail, Figure 4.12a presents a simplified drawing of the triaxial loading facility utilized by Siskind (1970) for tests on gas storage reservoir models containing a pressurized cavity. In these studies a PZT element located in the upper section of the loading piston (LP) was used to monitor AE/MS activity in the model. Figure 4.12b illustrates typical AE/MS signals detected in two models (No. 41 and 44) during the internal pressurization of the central cavity. These signals clearly illustrate the primary event (P) and series of secondary events (S$_1$, S$_2$, and S$_3$) due to reflections within the test facility. Based on a preliminary review of the facility geometry it appeared realistic to assume that the major reflection surfaces were the bottom of the lower loading head (LLH), which is partially acoustically isolated from the bottom plate (BP) by a rubber o-ring seal, and the bottom, interior surface of the head-section of the loading piston (LP) where the PZT element was attached. The following somewhat simplified analysis appears to validate this assumption.

1. Based on the data presented in Figure 4.12b the times between secondary signals are t$_1$ = 190 μs and t$_2$ = 180 μs giving an average reflection time t̄ = 185 μs.

2. The propagation path is assumed to be approximately 2L$_P$ + 2L$_S$ in length; however, the propagation velocity in the steel loading piston (C$_P$ = 5800 m/s) and in the model material (C$_M$ = 2500 m/s) are different.

3. Based on approximate values of L$_P$ = 0.273 m and L$_S$ = 0.060 m, the theoretical reflection time may be computed, namely;

$$t_t = \sum \frac{2d_i}{C_i} = \frac{2L_P}{C_P} + \frac{2L_s}{C_s} \qquad \text{(Eq. 4.3)}$$

$$= \frac{2 \times 0.273}{5800} + \frac{2 \times 0.064}{2500} = (94.1 + 51.2) \times 10^{-6}$$

$$t_t \approx 145 \mu s$$

4. The calculated reflection time (t$_t$ ≈ 145 μs) and average measured time between secondary signals t̄ ≈ 185μs are found to agree within ≈ 20 percent.

Considering the difficulty of accuracy determining the reflection time and the obviously complex acoustic path involved, the result appears to substantiate the fact that secondary events (S$_1$, S$_2$ and S$_3$) are due to reflections within the test vessel.

The preceding example indicates clearly the problems due to reflection associated with the indirect measurement of AE/MS activity. In this example the primary events appeared to have a duration of the order of 200 μs; however, due to the various secondary arrivals the effective duration of the received signal was often of the order of several milliseconds. Certainly in those experiments where the event rate was low, and as a result the average time between events was large compared with the duration of the effective signal (primary and multiple secondary events), manual analysis or some form of electronic processing (e.g., gating circuit) could be utilized to overcome this problem. However, at higher event rates there would be little possibility of separating primary and secondary (reflection) events and the output of any automatic system used to monitor the AE/MS data from such an experiment would be of limited value.

Based on the preceding discussions it is clear that in triaxial tests, where possible, AE/MS transducers should be mounted directly on the test specimen. In the past, piezoelectric elements, and to a limited degree semiconductor strain gages (see Khair, 1972), have been utilized. One of the major difficulties in this type of installation is the problem of making the necessary electrical connections to the transducers since they are located under the specimen jacket. Khair (1972) describes how this problem may be resolved. For his studies on large gas storage reservoir models a special triaxial vessel was designed in which a series of electrical connections were provided between a set of external connectors and an annular space located between the test specimen and the outer jacket. Prior to testing, the specimen was placed on the base of the triaxial vessel and all electrical connections are made. The outer jacket was then installed and the upper section of the triaxial vessel lowered and locked in position reading for testing.

A later paper by Byerlee and Lockner (1977) describes an alternate method. Here a relatively thick-walled (0.00475 m) polyurethane jacket was utilized to seal the confining pressure fluid from the test specimen. As illustrated in Figure 4.13, PZTY elements were attached to the ends of hardened steel plugs, the other ends of which were ground to the same radius of curvature as the test specimen. These plugs were attached to the specimen using epoxy cement, passed through holes cut in the jacket material, and sealed in place with epoxy cement. Electrical connections from the transducers were brought through suitable high pressure connections, located at one end of the triaxial vessel, to the associated AE/MS monitoring system.

Tests in hostile environments: The measurement of AE/MS activity in laboratory specimens or structures subjected to a hostile environment (e.g., high temperature, high magnetic or radiation fields, etc.) presents special problems due to the effect of these environments on the transducer itself. In such cases special transducers developed for these environments must be utilized or alternately the associated AE/MS activity must be transmitted out of the hostile environment by some form of "waveguide". Figure 4.11d, illustrates a typical example of the latter technique, which could be used, for example, to monitor AE/MS activity in a specimen being tested under high temperature conditions. Here a suitable metal or ceramic waveguide would be cemented to the specimen,

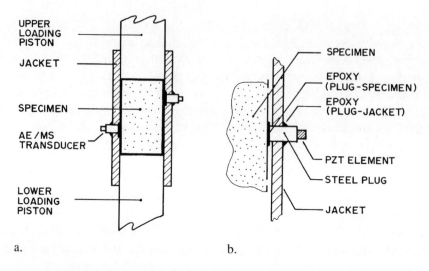

Figure 4.13. Details of an alternative method of installing AE/MS transducers for tests under triaxial loading (after Byerlee and Lockner, 1977). a. Overall view of loading configuration. b. Details of transducer installation.

projecting out through the wall of the test enclosure where a conventional AE/ MS transducer may be attached. Using such a technique care must be taken to prevent any contact between the waveguide and the enclosure, and to insure that consideration is given to the possible generation of stress wave reflections in the waveguide itself. Papers by Drew and Seitz (1981), Kumagai et al. (1981), and Shirawa et al. (1981) describe the use of the waveguide technique for high temperature AE/MS studies on rock-like materials such as ceramics and firebrick.

4.4 LABORATORY MONITORING FACILITIES

4.4.1 General

The basic format of AE/MS monitoring systems utilized in laboratory studies are similar to those described earlier for field use (see Chapter 3). Sophisticated hybrid systems have been utilized for some years in the study of non-geologic materials (e.g., metals, plastics, etc.) and for rock-like materials (e.g., concrete, ceramics, etc.). Such systems have been utilized in only limited cases for studies on geologic materials, however, their use in such studies has increased steadily since the early 1980s. For the most part, however, AE/MS monitoring facilities for use in laboratory tests on geologic materials have been of the basic- and parametric-type. It is also important to note, that due to the special character of AE/MS laboratory data (e.g., high frequency content, high event rates, etc.), monitoring facilities specifically designed for field use are generally not suitable for use in the laboratory.

4.4.2 Basic systems

Basic AE/MS monitoring systems for laboratory use have a similar format to those used in the field (see Fig. 3.17). However, the transducer, and the signal conditioning and readout sections must have the characteristics necessary to insure that the higher frequencies and higher event rates associated with laboratory studies can be accommodated. In this regard the most sensitive sections of the monitoring system, besides the transducer itself, is the readout section. Whereas mechanical or lightbeam oscillographic recorders and conventional magnetic tape recorders may be utilized for recording of field data, they are generally not satisfactory for laboratory purposes.* As a result only high quality, high speed magnetic tape recorders or some form of digital storage are practical. Such facilities are generally expensive and as a result the use of basic-type monitoring systems is relatively limited.

Figure 4.14 presents a block diagram of a typical basic-type laboratory system. Systems of this general format were utilized during many of the early studies carried out at Penn State (e.g., Chugh, 1968; Kim, 1971; Siskind, 1970) and elsewhere. In such an arrangement the output of the signal conditioning system is recorded on a suitable analog magnetic tape recorder, and signals are often displayed on a conventional or storage-type oscilloscope (CRO). In those cases where signal frequencies are sufficiently low (f < 15 kHz) an audible output is often provided from the conditioned signal using a suitable audio amplifier and speaker. When high frequency signals (f > 15 kHz) are involved it is necessary to incorporate a "heterodyne circuit" prior to the audio amplifier in order to artificially reduce the actual signal frequency to the lower audio range. Experience has indicated that audible output signals are often useful for manual separation of various types of AE/MS signals and for detection of background noise.

Laboratory-type multi-channel basic systems, having the same general format as the single channel system shown in Figure 4.14, have also been employed where source location data is required. In such cases a single visual (CRO) and audio monitor are employed with a suitable switching arrangement so that any of the channels may be selectively monitored. Rothman (1975), for example, utilized the Penn State field monitoring facility shown earlier in Figure 3.19 for this purpose. It should be pointed out that this was only possible due to the fact that the facility was specifically designed to have the capability of recording AE/MS signals up to a frequency of approximately

---

* While a number of early AE/MS laboratory studies were carried out at relatively low frequencies (i.e., 100 Hz – 50 Hz), most recent studies have been in the range 50 kHz – 500 kHz where the majority of the source energy appears to be concentrated.

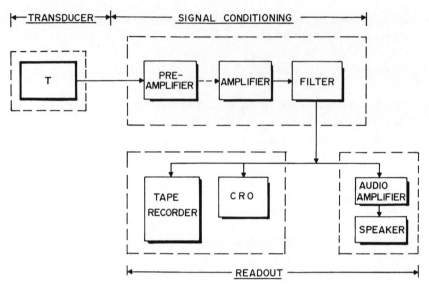

Figure 4.14. Block diagram of a typical basic-type AE/MS laboratory monitoring system.

500 kHz. Conventional field monitoring systems would not have such a high frequency capability.

4.4.3   Parametric systems

The most common AE/MS monitoring system for laboratory use is the single channel parametric-type. The block diagram of one such parametric system has been illustrated earlier in Figures 3.18 and 4.15 and indicates the general form of such systems. As indicated earlier the major disadvantage of a parametric system, at least in field applications, is the fact that the detailed character of the AE/MS signals are lost. For laboratory studies, however, this disadvantage may be somewhat less important due to the fact that such studies are often of a more routine nature. Furthermore, such a system, by virtue of its parametric format is capable of conveniently processing the large number of high frequency signals normally encountered.

As illustrated in Figure 4.15 the laboratory-type parametric system incorporates a two-part signal conditioning section with the first part (analog section) being of conventional design. Analog output signals from this section (one or more channels) are fed into the second section (parametric section) where the desired AE/MS parameters are generated and stored for subsequent output to various types of readout devices.

Since the parameter development section consists of a number of elements combined in various ways to provide the desired AE/MS parameters a number of such elements will be briefly considered at this point. These include, the

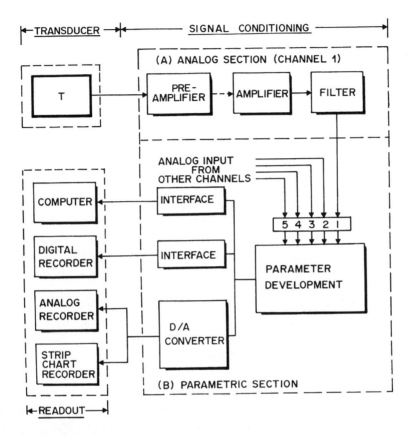

Figure 4.15. General form of a parametric-type AE/MS laboratory monitoring system.

threshold detector, envelope processor, time gate, counter, etc. In the present context, the functions and limitations of these elements will be given major consideration rather than the circuits themselves, and as a result they will be considered as "black boxes". Those interested in the details of the theory and design of the associated circuits should refer to electronics texts dealing with pulse and digital concepts (e.g., Bartee, 1966; Cerni and Foster, 1962; Malmstadt et al., 1963; Wolfe, 1983).

Threshold detector: The purpose of a threshold detector (TD) is to convert analog AE/MS signals to a pulse format. Figure 4.16 illustrates an incoming analog signal and the equivalent pulse output based on a threshold level of Δv. The threshold detector is normally some form of Schmidt trigger circuit (Malmstadt et al., 1963) which produces a square pulse of fixed amplitude and duration each time the incoming signal exceeds a specific voltage level, denoted as the threshold level.

As noted in Figure 4.16, the incoming signal exceeds the threshold level Δv three times (points 1, 3 and 5) during the duration of the event (ΔT). At each of these times a pulse is generated at the output of the TD. In AE/MS nomenclature these are known as ring-down pulses since they occur during the decay, or ring-down, of the incoming signal. It is important to note that in this case a single event generates three ring-down pulses. In general, AE/MS activity may therefore be described in terms of the number of event counts (N_E) or the number of ring-down counts (N_{RD}). This presents a certain degree of ambiguity since the relationship between N_{RD} and N_E will depend on the amplitude of the incoming event and the selected threshold level.

A preliminary consideration of the problem would suggest that, as illustrated in Figure 4.17a, the ambiguity could be removed by raising the threshold level from V_1 to V_2, since at the V_2 level $N_{RD} = N_E$. However, Figure 4.17b illustrates the situation where two legitimate events of different amplitude are involved. In this situation, at a threshold level of V_2, a single ring-down pulse (I_i') is obtained for event 1 but none is obtained for event 2. Furthermore, when the threshold level is lowered sufficiently to provide a single ring-down pulse for event 2 (I_2) two such pulses are generated for event 1 (I_1 & I_2). Two common techniques for resolving the problem, namely, the use of an envelope processor or a time gate are discussed briefly in the following sections.

Envelope processor: One method for establishing the relationship $N_{RD} = N_E$ is to utilize an envelope processor unit. As illustrated in Figure 4.18, this unit is inserted in front of the threshold detector and processes the event signal (A) to provide an output equivalent to the envelope (B) of the signal. Such a signal, when passed into the threshold detector, will generate a single output pulse for each input event. Such an envelope processor is basically a rectifier circuit involving capacitance-resistance circuits and their associated charge-discharge characteristics. The envelope processor approach has been found to be satisfactory in many cases; however, it is limited by the time constants of

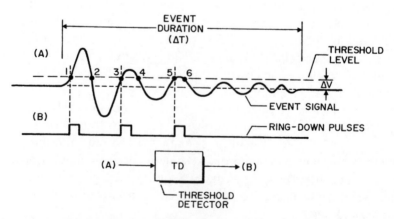

Figure 4.16. Pulse output from a threshold detector in response to an incoming AE/MS event. ((A) – incoming AE/MS event, (B) – pulse output.)

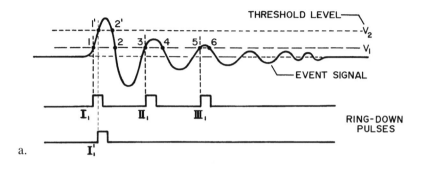

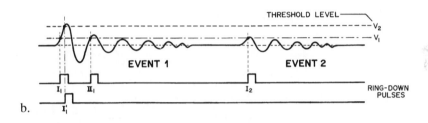

Figure 4.17. Effects of different threshold levels on the output of a threshold detector. a. For events of similar amplitude a unique threshold level may be selected to insure that $N_{RD} = N_E$. b. For events of varying amplitude a unique threshold level to insure $N_{RD} = N_E$ is not possible.

the rectifier circuit. For a selected set of rectifier circuit components it is only efficient over a specific range of input signal frequencies.

Time gate unit: A second technique for obtaining event pulse counts rather than ring-down counts from a threshold detector is to employ a gating circuit similar to that shown in Figure 4.19b. Figure 4.19a illustrates a typical sequence of AE/MS event signals having an event duration of ΔT and a time between events of TT. In general the time gate technique is only applicable when $\Delta T <$ TT.

Using the time gate unit the AE/MS signal (A) is first fed into a threshold detector generating a sequence of ring-down pulses (B). These are fed into a gating circuit which generates a single output pulse (C) then remains closed for a preset gating time (T_G). By adjusting this gating time to a value of $\Delta T < T_G <$ TT only one output pulse will be generated for each AE/MS event. This technique has been utilized successfully during a number of studies (e.g., Hardy, 1972; Siskind, 1970). It is important to note, however, that the technique is sensitive to the values of ΔT and TT which will depend on the type of material under study and the associated test conditions.

Floating threshold detector: As illustrated in Figure 4.20a, the threshold detector level is normally established above the background noise, which is

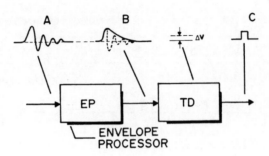

Figure 4.18. Use of an envelope processor to insure $N_{RD} = N_E$. ((A) – incoming AE/MS event, (B) rectified signal, (C) – Pulse output.)

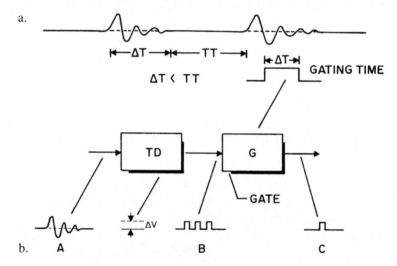

Figure 4.19. Use of a time gate to insure $N_{RD} = N_E$. a. Input event. b. Circuit operation. ((A) – incoming AE/MS event, (B) associated, ring-down pulses, (C) – pulse output.)

assumed to occur at a relatively constant level. This prevents the background noise level from generating ring-down counts. However, as noted in Figure 4.20b, if there are major variations in the background noise the fixed threshold level may on occasions be exceeded, generating erroneous counts. If a floating threshold detector is utilized the actual threshold level will be the sum of the peak voltage associated with the background noise, and the established fixed threshold voltage setting. As illustrated in Figure 4.20b the floating threshold avoids the problem of erroneous counts being generated when varying level background noise exceeds the fixed threshold level. It should be noted, however, that the fixed threshold method should be employed in all cases where the background noise level does not vary appreciably.

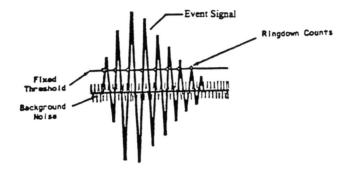

a.

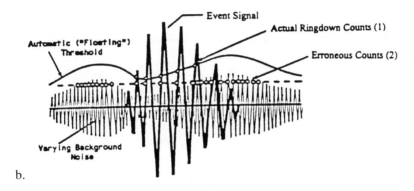

b.

Figure 4.20. Comparison of fixed and floating threshold methods (after Anon. 1981). a. Fixed threshold method. b. Floating threshold method. ((1) & (2) illustrate, respectively, output counts with floating threshold enabled and disabled.)

Digital counter: The function of a digital counter, denoted here simply as a counter, is to accept incoming pulses, and to count, totalizer and store these. As illustrated in Figure 4.21, stored data may be displayed visually using accessory elements, and such counters are usually equipped with circuits to provide start, stop and reset (storage dump) functions.

Clock-pulse generator: The function of a clock-pulse generator is to provide a continuous output of uniform pulses spaced at a specific period of time (constant repetition rate) for various timing purposes. A clock-pulse generator usually contains a highly stable quartz crystal oscillator which generates a sinusoidal signal of specific frequency. This signal is applied to the input of a rate selection circuit, normally involving a series of so-called digital dividers, which in turn provides a signal of selectable frequency to a pulse generator. The output of this generator are the required clock-pulses. Figure 4.22a illustrates a simple block diagram of a typical clock-pulse generator.

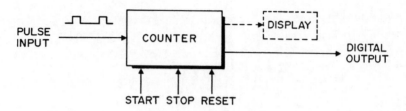

Figure 4.21. Format of a typical digital counter.

Triggered gate: A triggered gate is a gating circuit which is controlled by an external trigger pulse. Figure 4.22c illustrates a typical representation for such a circuit. In operation, application of certain types of trigger pulses either allow or prevent input data from passing through the gate.

RMS module: In some situations the AE/MS signals involve a large number of superimposed events or appear to be more-or-less continuous. In such cases the use of ring-down or event counts becomes impracticable. Past experience has indicated that such signals are best analyzed statistically, and a root-mean (RMS) model is sometimes employed. As shown in Figure 4.22b in such a model the output voltage represents the RMS level of the AE/MS signal input.

The various components described in the previous sections may be combined in a variety of forms. However, at this stage in the text major consideration will be given to three common forms of parameter development, namely, those used to provide an accumulated or total count, count rate, and count rate-memory data. In all cases it is assumed that the associated threshold detector provides a ring-down pulse output.

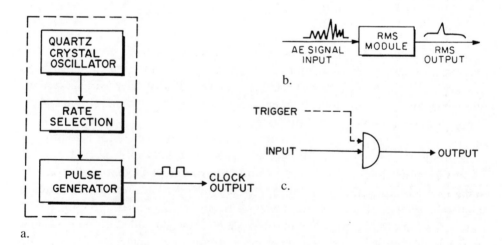

Figure 4.22. Additional elements utilized in AE/MS parametric systems. a. Clock-pulse generator. b. RMS module. c. Triggered gate.

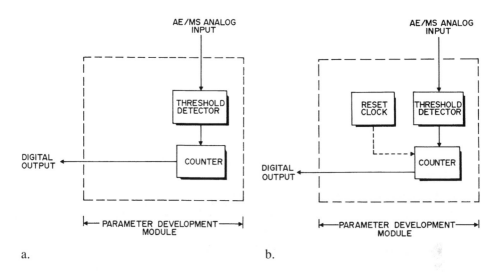

Figure 4.23. Block diagrams for two types of parameter development module. a. Generating accumulated (total) count data. b. Generating count rate data.

Total (accumulated) count module: Figure 4.23a illustrates the block diagram for a parameter development module for generating accumulated or total count data. Here pulses generated by the threshold detector, in response to the AE/MS analog input signal, are fed into a counter. The output of the counter is a digital signal equivalent to the total number of accounts accumulated.

Count rate module: In order to generate count rate data the circuit shown earlier in Figure 4.23a is modified, as illustrated in Figure 4.23b, by the addition of a reset clock. The purpose of the reset clock is to provide a reset pulse to the counter at selected intervals of time. In operation, pulses generated by the threshold detector, in response to the incoming signals, are fed into the counter and accumulate until a reset pulse is generated by the clock. This dumps the counter, dropping the digital output to zero, and initiates a new counting sequence. The equivalent count rate of such a system is equal to the total number of counts accumulated between reset pulses divided by the time between such pulses. Figure 4.26c, presented later, illustrates the analog output of a parametric system operating in the count rate mode.

Count rate-memory module: One disadvantage of the standard count rate system is the fact that each time the counter is reset the output of the monitoring system drops suddenly to zero. This results in the sawtooth pattern are shown later in Figure 4.25c. The parameter development circuit shown in Figure 4.24, denoted as a count rate-memory circuit, which includes a digital memory and additional logic circuits, was designed to overcome the disadvantages of the simple count rate system.

During operation of the circuit shown in Figure 4.24, analog AE/MS signals are applied to the input of the threshold detector which provides an equivalent series of pulses to the counter. The counter accumulates the incoming pulses until it receives a suitable set of reset pulses. In this mode of operation the actual reset pulse is proceeded by a "strobe" pulse which transfers the counter data to a digital memory. Following this, the counter is reset to zero and the next counting sequence is initiated.

The digital output of the memory, which is held constant until updated by the next reset sequence, is equivalent to the number of counts accumulated between reset pulses. The count rate is therefore computed on the basis of the number of counts in memory divided by the time between reset pulses.

In order to better appreciate the response of the different types of parameter development modules, Figure 4.25 illustrates the analog output of parametric monitoring systems incorporating each of the three types of modules described in the preceding sections. The figure illustrates the effect of three different AE/MS input signal rates, namely, 10, 20 and 50 pulses per second (pps), each of which are held constant for a period of 60 s.

During AE/MS laboratory studies each of the different parameters (total counts, count rate and count rate-memory) offer certain advantages and disadvantages. When relatively few events are involved it is often convenient to utilize total counts. However, when high event rates are involved, or when long-term tests are involved, the counter may become overloaded, and count rate or count rate-memory should be utilized. In such situations count rate-memory is often the most suitable since it eliminates the large recorder pen

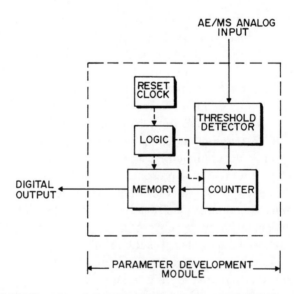

Figure 4.24. Block diagram of a parameter development module for generating count rate-memory data.

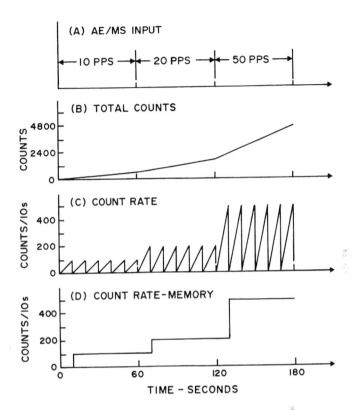

Figure 4.25. Analog output for three different types of parameter development modules in response to a specific AE/MS input. ((A) – AE/MS input; (B), (C), (D) – response of monitoring system containing total count, count rate, and count rate-memory modules.)

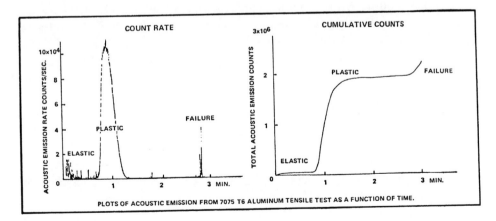

Figure 4.26. AE/MS data from 7075-T6 aluminum tensile test as a function of time presented in terms of count rate and total [Cumulative] counts (after Anon., 1983).

excursions which occur during counter reset when the count rate mode is employed. Care must be taken, however, in correlating count rate-memory with other experimental parameters (e.g., applied stress or specimen strain), since the rate data displayed is based on data accumulated during the previous reset period.

Figure 4.26 illustrates clearly the type of data presentation obtained using two different methods of parametric processing. In this figure the AE/MS data obtained during a tensile test on 7075-T6 aluminum are displayed in terms of count rate and total (cumulative) count. In the count rate mode the elastic, plastic and failure regions are clearly indicated. In contrast, although the cumulative count presentation does not emphasize the various phases of deformation and failure, the data are seen to provide a curve which closely resembles the typical stress-strain curve obtained in such a test.

Over the last 25 years commercial firms have developed and marketed instrumentation for monitoring and processing of high frequency AE-data. Parametric-type systems (see Fig. 4.15) marketed by a number of firms including, Acoustic Emission Technology, Brüel & Kjaer Instruments, Dunegan Corporation and Physical Acoustics Corporation, have been used in laboratory and special field studies associated with a variety of geotechnical applications. Photographs of a number of these instruments are shown in Figure 4.27.

Figure 4.28 illustrates the basic format of the model 204GR miniature acoustic emission system marketed by Acoustic Emission Technology. The 204GR is a single channel facility capable of operating on a built-in rechargeable battery supply. This makes it particularly useful for field studies. It was designed to supply a variety of outputs including AE signals, event pulse, RMS signal level, as well as a recorder output for plotting total events/counts or event/count rate. It also includes the capability of utilizing a fixed or a "floating" AE signal threshold.

Another unique AE/MS monitoring facility is the 3000 series system developed by the Dunegan Corporation. Figure 4.29 illustrates the basic format of the system. It was built on the modular concept, incorporating an internal data bus arrangement allowing for the addition of a series of special purpose models. Modules available included a spatial discriminator, amplitude detector and a distribution analyzer.

In the early 1970s microprocessor technology rapidly developed and new forms of parametric AE/MS monitoring systems started becoming available. These systems in many cases utilized traditional hard-wired digital front ends, but added microprocessor units for the storage and processing of parametric data. Multichanel systems such as the AET 5000 system, manufactured by Acoustic Emission Technology, and the four-channel model 3104 system and the SPARTAN system with 128 channel capacity, manufactured by the Physical Acoustics Corporation are examples of the advances in AE/MS system development.

a.

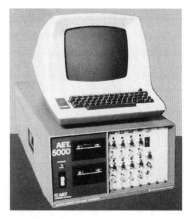

c.

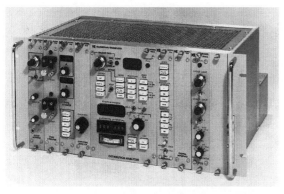

b.

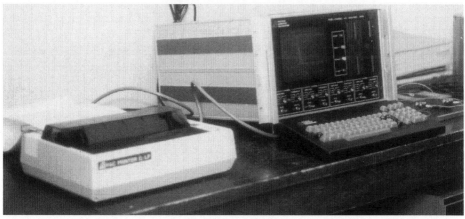

d.

Figure 4.27. Photographs showing a number of parametric-type AE/MS monitoring systems. a. AET model 204 GR. b. Dunegan 3000 system. c. AET model 5000 system. d. PAC model 3104 system.

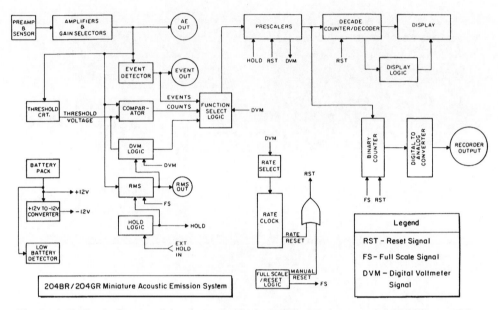

Figure 4.28. Basic format of the Acoustic Emission Technology model 204GR portable, single channel AE/MS monitoring system (after Anon., 1981)

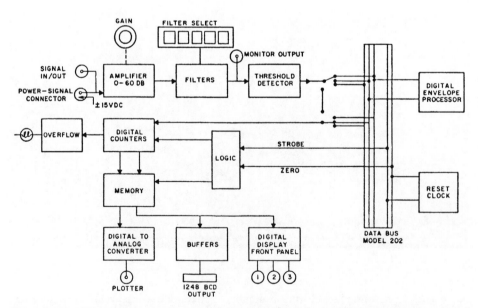

Figure 4.29. Basic format of the earlier Dunegan 3000 series AE/MS monitoring system (after Anon., 1983).

In the late 1980s, researchers and equipment manufacturers became increasingly concerned relative to the current state-of-the-art in AE/MS systems and the direction of future development. A number of import papers from this era include those by Hamstad (1988), Mobley (1987), Oyaizu et al. (1991), Vahaviolos (1986), and Yamaguchi (1988).

In the last few years there has been another major step in high frequency AE/MS instrumentation. Here most of the analog and the hard-wired digital circuitry has been replaced by a computer-based acquisition, storage and processing system. The new systems are similar in form to that denoted, earlier in the chapter on field monitoring as hybrid/digital systems (see Fig. 3.23). In many of these new systems, even the filters are digital in form and such factors as amplifier gains are set through the associated soft-wave. Furthermore, the soft-ware allows for sophisticated statistical analysis of data and comprehensive on-screen graphical displays. Firms such as Digital Wave, Dunegan Engineering Consultants, Physical Acoustics, and Vallen-Systems GmbH have developed systems of this type. The Vallen model AMSY4 multi-channel system, shown in Figure 4.30, is an excellent example of this sophisticated type of instrumentation. A number of systems have been developed so they may also be utilized to carry out, so-called, "AE Model Analysis" (Gorman, 1994). Further details on this technique are included later under special topics.

Besides the systems specifically manufactured for AE/MS laboratory studies, a number of manufacturers, including R. C. Electronics Inc., California, and Gage Applied Sciences, Inc., Montreal, Canada, supply multi-channel A/D

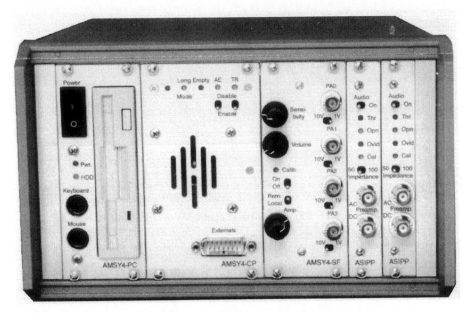

Figure 4.30. Vallen model AMSY4 computer-based monitoring system.

and signal processing cards for PC application along with associated software. These so-called "computerscope"/ "compuscope" facilities may be installed in a suitable personal computer to provide AE/MS signal acquisition, storage and processing. The author has had considerable success with the R. C. Electronics, IS-16 data acquisition system card ("COMPUTERSCOPE") installed in a typical 386 PC. This system is capable of digitizing and recording data from up to 16 channels. Using a single channel, an A/D rate of 1 MHZ is possible. A variety of data analysis software is available for data storage upgrade, data conditioning, signal averaging and power spectrum analysis. This system has been successfully utilized by the author in a number of studies (e.g. Shen, Hardy and Khair, 1997). Much higher A/D rate systems are currently available. For example, Gage Applied Sciences markets the Compuscope 225 which allows two channel operation at a rate of 25 MHZ for both channels.

4.4.4 Readout systems

Various types of readout systems are utilized for displaying and/or storing AE/MS laboratory data. For example, systems like the Dunegan Model 3000 series and the AET Model 204GR, shown in block diagram form in Figures 4.28 and 4.29, have a built-in digital display, an analog output for operation of a plotter or strip chart recorder, direct output of a number of AE parameters and in some cases a digital output for connection to an external computer. The more sophisticated multi-channel parametric systems utilize built-in computers and store data on floppy disc for subsequent processing and output to digital printers and plotters. In contrast, the latest state-of-the-art systems are capable of recording and playback of full waveform data, detailed signal processing and highly detailed on-screen presentations.

Laboratory studies of AE/MS activity in geologic materials present a number of unique problems. For example, experiments are often carried out over extended periods of time and relatively high event/count rates are often experienced. During some experiments very high speed multi-channel analog recorders have been utilized. Such high tape speeds are necessary to enable recording of signals in the range of 100 kHz – 300 kHz. For example, Rothman (1977), used a Sangamo Model 3614 magnetic tape recorder, operating at a tape speed of 60 ips ($\approx$ 150 cm/sec) to record multi-channel AE/MS signals during creep experiments on a fine grained sandstone (Tennessee Sandstone). Based on the speed used and the maximum available sized tape, a total recording time of less than 20 minutes was possible.

An alternative system, used by the author incorporates a modified reel-to-reel, Sony video tape recorder. With accessory equipment, the video recorder is capable of continuous recording of high frequency AE/MS signals for periods of up to one hour. The only disadvantage of this arrangement is the relatively low dynamic range of video recorder, < 30 dB, which limits the range of signal

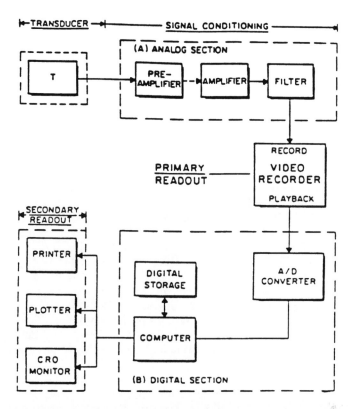

Figure 4.31. Typical arrangement for single-channel on-line recording using a modified video recorder and subsequent playback and digital recording, processing and display.

amplitudes that may be recorded. Figure 4.31 illustrates a typical arrangement for single-channel on-line recording using a modified video recorder. A high speed analog recorder could also be utilized in this application but the maximum recording time would be severely reduced. In the system shown the video or analog recorder provides the primary readout (data storage). Off-line, selected sections of the recorded data may be played into the digital section of the facility for digital recording, processing, and display on the secondary readout system.

REFERENCES

Anon. 1980. *Electronic Instruments – Master Catalog 1980*, Bruel and Kjaer Instruments, Inc., Marlborough, Massachusetts.

Anon. 1981. *Operating Instructions for the AET Models 204BR and 204GR Miniature Battery-Powered Acoustic Emission Systems*, Acoustic Emission Technology Corporation, Sacramento, California.

Anon. 1983. Personal Communication, Dunegan/Endevco, San Juan Capistrano, California, January 1983.

Bartee, T.C. 1966. *Digital Computer Fundamentals*, McGraw-Hill Book Company, New York.

Byerlee, J.D. and Lockner, D. 1977. Acoustic Emission During Fluid Injection, *Proceedings First Conference on Acoustic Emission/Microseismic Activity in Geologic Structures and Materials*, The Pennsylvania State University, June 1975, Trans Tech Publications, Clausthal, Germany, pp. 87-98.

Cerni, R.H. & Foster, L.E. 1962. *Instrumentation for Engineering Measurement*, John Wiley & Sons, Inc., New York.

Chugh, Y.P. 1968. *An Investigation of Frequency Spectra of Microseisms Emitted from Rock Under Tension in the Range 300-15,000 cps*, M.S. Thesis, Geomechanics Section, Department of Mineral Engineering, The Pennsylvania State University.

Doebelin, E.O. 1975. *Measurement Systems*, McGraw-Hill Book Company, New York.

Dove, R.C. and Adams, P.H. 1964. *Experimental Stress Analysis and Motion Measurement*, Charles E. Merill Books, Inc., Columbus, Ohio.

Drew, J.M. and Seitz, M.G. 1981. Applications of Acoustic Emission Analysis with Industrial Ceramics, *Advances in Acoustic Emission*, H. L. Dunegan and W. F. Hartman (eds.), Dunhart publishers, Knoxville, pp. 204-212.

Eitzen, D. and Breckenridge, F.R. 1987. Acoustic Emission Sensors and Their Calibration, *Nondestructive Testing Handbook*, Vol. 5: Acoustic Emission Testing, Section 5, Editor: P. M. McIntire, American Society for Nondestructive Testing, Columbus, Ohio, pp. 121-134.

Glaser, S.D. and Nelson, P.P. 1992a. High-Fidelity Waveform Detection for Acoustic Emissions From Rock, *Materials Evaluation, Vol. 30*, No. 3, pp. 354-366.

Glaser, S.D. and Nelson, P.P. 1992b. Acoustic Emissions Produced by Discrete Fracture in Rock–Part 2: Kinematics of Crack Growth During Controlled Mode I and Mode II Loading of Rock, *International Journal of Rock Mechanics, Vol. 29*, pp. 253-265.

Gorman, M.R. 1994. New Technology for Wave Based Acoustic Emission and Acousto-Ultrasonics, *Wave Propagation and Emerging Technologies*, ASME Publication, AMD-Vol. 188, ASME, New York, pp. 47-59.

Hamstad, M.A. 1988. Characterization and Measurement Accuracy of Acoustic Emission Systems, *Progress in Acoustic Emission*, Proceedings of the International Acoustic Emission Symposium, Kobe 1988, K. Yamaguchi, I. Kimpara and Y. Higo (eds.), Japanese Society for Non-Destructive Inspection, Tokyo, pp. 121-132.

Hardy, H.R. Jr. 1972. *A Study to Evaluate the Stability of Underground Gas Storage Reservoirs*, AGA Catalog No. L 19725, American Gas Association, Arlington, Virginia.

Khair, A.W. 1972. *Failure Criteria Applicable to Pressurized Cavities in Geologic Materials Under In-Situ Stress Conditions*, Ph.D. Thesis, Geomechanics Section, Department of Mineral Engineering, The Pennsylvania State University.

Kim, R.Y. 1971. *An Experimental Investigation of Creep and Microseismic Phenomena in Geologic materials*, Ph.D. Thesis, Geomechanics Section, Department of Mineral Engineering, The Pennsylvania State University.

Kimble, E.J. Jr. 1984. Semiconductor Strain Gages as Acoustic Emission Transducers, *Proceedings Third Conference on Acoustic Emission/Microseismic Activity in Geologic Structures and Materials*, The Pennsylvania State University, October 1981, Trans Tech Publications, Clausthal-Zellerfeld, Germany.

Kumagi, M., Uchimura, R. and Kisidaka, H. 1981. Studies of Fracture Behavior of Refractories Under Thermal Shock Conditions Using an Acoustic Emission Technique, *Advances in Acoustic Emission*, H.L. Dunegan and W.F. Hartman (eds.), Dunhart Publishers, Knoxville, pp. 233-248.

Malmstadt, H.V., Enke, C.G. and Toren, E.C. Jr. 1963. *Electronics for Scientists*, W.A. Benjamin, Inc., New York.

Mobley, K. et al. 1987. Survey of Commercial Sensors and Systems for Acoustic Emission Testing, *Nondestructive Testing Handbook*, Vol. 5: Acoustic Emission Testing Section, Editor: P. M. McIntire, American Society for Nondestructive Testing, Columbus, Ohio, pp. 513-549.

Norton, H.N. 1969. *Handbook of Transducers for Electronic Measuring Systems*, Prentice Hall, Inc., Englewood Cliffs, New Jersey.

Ookouchi, T., Itou, M. and Isaji, H. 1991. AE Signal Transmission System from Rotating Sensor to Stationary Field, *Proceedings of 4th World Meeting on Acoustic Emission and 1st International Conference on Acoustic Emission in Manufacturing*, September 1991, Boston, Extended Paper Summaries and Abstracts, pp. 527-534.

Oyaizu, H., Yamaguchi, K. and Jiang, W. 1991. An Approach to Evaluation for AE Instrumentation, *Proceedings of 4th World Meeting on Acoustic Emission and 1st International Conference on Acoustic Emission in Manufacturing*, September 1991, Boston, Extended Paper Summaries and Abstracts, pp. 511-518.

Proctor, T.M. 1982a. An Improved Piezoelectric Acoustic Emission Transducer, *Journal of the Acoustical Society of America, Vol. 71*, No. 5, pp. 1163-1168.

Proctor, T.M. 1982b. Some Details on the NBS Conical Transducer, *Journal of Acoustic Emission, Vol. 1*, No. 3, pp. 173-178.

Proctor, T.M. 1986. More-Recent Improvements on the NBS Conical Transducer, *Journal of Acoustic Emission, Vol. 5*, No. 4, pp. 134-142.

Richardson, A.M. 1978. *An Experimental Investigation of the Uniaxial Yield Point of Salt*, M.S. Thesis, Geomechanics Section, Department of Mineral Engineering, The Pennsylvania State University.

Roberts, D.A. 1981. *An Experimental Study of Creep and Microseismic Behavior in Salt*, M.S. Thesis, Geomechanics Section, Department of Mineral Engineering, The Pennsylvania State University.

Rothman, R.L. 1975. *Crack Distribution Under Unixial Load and Associated Changes in Seismic Velocities Prior to Failure*, Ph.D. Thesis, Geophysics Section, Department of Geosciences, The Pennsylvania State University.

Rothman, R.L. 1977. Acoustic Emission in Rock Stressed to Failure, *Proceedings First Conference on Acoustic Emission/Microseismic Activity in Geologic Structures and Materials*, The Pennsylvania State University, June 1975, Trans Tech Publications, Clausthal-Zellerfeld, Germany, pp. 109-133.

Shen, W.H., Hardy, H.R. Jr. and Khair, A.W. 1997. Laboratory Study of Acoustic Emission and Particle Size Distribution During Rotary Cutting, *Int. J. Rock Mech. & Min. Sci., Vol. 34*, No. 3-4, paper 121, 12 pp.

Shirawa, T., Sakamoto, Y., Yamaguchi, H., Suzuki, T., Fujisawa, K. and Arabori, T. 1981. Acoustic Emission Characteristics of Firebricks, in *Advances in Acoustic Emission*, H.L. Dunegan, and W.F. Hartman (eds.), Dunhart Publishers, Knoxville, pp. 213-232.

Siskind, D.E. 1970. *The Pressurization and Failure of Model Underground Openings*, Ph.D. Thesis, Geophysics Section, Department of Geosciences, The Pennsylvania State University.

Vahaviolas, S.J., 1986, 3rd Generation AE Instrument Techniques for High Fidelity and Speed of Data Acquisition, *Progress in Acoustic Emission III*, Proceedings 8th International Acoustic Emission Symposium, Tokyo 1986, K. Yamaguchi, A. Aoki and T. Kishi (eds.), Japanese Society for Non-Destructive Inspection, Tokyo, pp. 102-116.

Weiss, G., and Glaser, S.D. 1998. Embeddible Sensor for Monitoring Hydrofrac Body Waves, *Proceedings Sixth Conference on Acoustic Emission/Microseismic Activity in Geologic Structures and Materials*, The Pennsylvania State University, June 1996, Trans Tech Publications, Clausthal-Zellerfeld, Germany, pp. 489-503.

Wolf, S. 1983. *Guide to Electronic Measurements and Laboratory Practice*, Second Edition, Prentice-Hall, Inc., Englewood Cliffs, New Jersey.

Yamaguchi, K. 1988. Instrumentation and Data Processing for Acoustic Emission Technology and Applications, *Progress in Acoustic Emission IV*, Proceedings 9th International Acoustic Emission Symposium, Kobe 1988, Editors: K. Yamaguchi, I. Kimpara and Y. Higo, Japanese Society for Non-Destructive Inspection, Tokyo, pp. 1-10.

CHAPTER 5

Source location techniques

5.1 INTRODUCTION

5.1.1 General

As noted earlier, one of the major advantages of the AE/MS technique is that, by analysis of data from a suitable transducer array, one can locate the source of the observed instability. Two techniques are commonly used, denoted in the geophysics literature as "single station" and "multiple station" seismometry. In the former, a single station, incorporating a triaxial transducer is utilized. The latter employs a number of single transducer stations. The multiple station technique is most commonly used in both AE/MS field and laboratory studies and will be discussed in detail. Single station and hybrid techniques will be considered further later in this chapter.

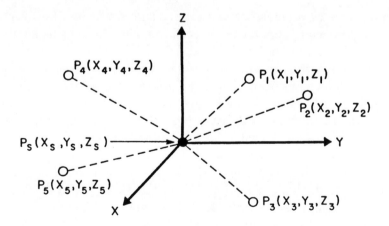

Figure 5.1. Geometry of a typical AE/MS transducer array. (Transducers are located at points P_1. —, P_5 with the AE/MS source at point P_s.)

Figure 5.1 illustrates a typical multiple station array containing five transducers, surrounding an AE/MS source, which is assumed to be located at point P_s. Various methods of source location analysis are presently available and will be briefly considered.

Multiple transducer source location methods employ a number of single transducer stations. AE/MS signals received by the various transducers are processed to provide such parameters as P- and S-Wave arrival times, arrival time sequence, signal amplitude, etc. The most common analysis methods used are the direct and the least squares travel-time-difference methods. Other analysis procedures include zonal and arrival-time-sequence methods, cross correlation techniques, and the use of cube stepping and simplex algorithms. These analysis procedures will be discussed further, later in this chapter.

5.1.2   Travel-time-difference method

An analytical technique, generally known as the travel-time-difference method, is commonly employed for multiple station source location analysis. Computerized methods for determining source locations normally utilize a least-squares iteration technique which searches for the best average solution for a set of equations which contain the transducer coordinates, x_i, y_i and z_i ($i = 1,2,3,...,n$), and the corresponding AE/MS arrival-times and velocities, t_i and V_i. The method involves the solution of the following set of i equations:

$$(x_i - x_s)^2 + (y_i - y_s)^2 + (z_i - z_s)^2 = V_i^2 (t_i - T_s)^2 \qquad \text{(Eq. 5.1)}$$

where x_s, y_s, and z_s are the true coordinates of the source, T_s is the true origin time, and the other parameters have been defined earlier. As there are four unknowns, x_s, y_s, z_s, and T_s, at least four equations (and therefore, data from four transducers) are necessary to solve the equations exactly. However, due to inherent experimental errors related to each equation (e.g., errors in transducer locations, determination of arrival-time differences, and velocities), at least five equations (data from five transducer stations) should be employed in order that the set of equations defined by Equation 5.1 are over-determined. This allows the computer program to minimize such error terms and to arrive at a best fit or average solution for the equation set. Additional equations (>5) will usually yield an even better value for the average solution because of further redundancy in the associated equation.

In general, the fundamental concept of AE/MS source location is based on geometry and the laws of motion. Figure 5.2 illustrates an AE/MS source location at the point P_s and a transducer at the point P_i. Most location theories assume that the stress wave originating at P_s travels via a straight line path to the point P_i where it is detected by the transducer. The time (t_i) for the stress wave to travel between these two points depends on the distance (d_i) and the

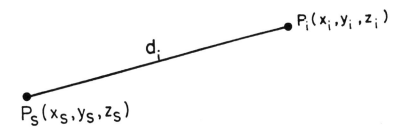

Figure 5.2. Relationships between source at P_s and transducers at P_i.

appropriate wave velocity. In three dimensional structures p- and s-wave data and their associated velocities are utilized, with p-wave data being the most common. In thin-walled structures or in near-surface geotechnical studies Rayleigh and Love wave data may also be utilized.

Referring to Figure 5.2 it was clear that the distance d_i may be described mathematically in two distinct ways, first in terms of the geometry (i.e., the coordinates of the two points) and second in terms of the time (t_i') required by the stress wave to travel between the two points. These two descriptions are represented by the following two equations,

$$d_i = \pm \sqrt{(x_i - x_s)^2 + (y_i - y_s)^2 + (z_i - z_s)^2} \qquad \text{(Eq. 5.2)}$$

and

$$d_i = V_i t_i' \qquad \text{(Eq. 5.3)}$$

where V_i is the appropriate stress wave velocity and the remaining parameters have been defined earlier. The general source location relationships are then obtained by equating the two expressions, namely:

$$\pm \sqrt{(x_i - x_s)^2 + (y_i - y_s)^2 + (z_i - z_s)^2} = V_i t_i' \qquad \text{(Eq. 5.4)}$$

Since all source location techniques involve a number of transducers (n) located in a suitable array, Equation 5.4 represents a series of n-equations which must be solved simultaneously to provide the values of the source coordinates (x_s, y_s and z_s).

An additional unknown is also involved in Equation 5.4, namely the time (t_i') required for the stress wave to reach the various transducers. In practice, although the transducers do provide information in regard to the time of arrival of the stress wave, the time at which this wave originates at the source (source time, T_s) is unknown. In general therefore

$$t_i' = t_i - T_s \qquad \text{(Eq. 5.5)}$$

where (t'_i) is the actual time, t_i is the measured time of arrival, and T_s is the source time. Substituting Equation 5.5 into Equation 5.4 yields

$$\pm \sqrt{(x_i - x_s)^2 + (y_i - y_s)^2 + (z_i - z_s)^2} = V_i (t_i - T_s) \qquad \text{(Eq. 5.6)}$$

As a result, data from a minimum of four transducers are required for three-dimensional source location, with the unknowns being x_s, y_s, z_s and T_s.

A considerable volume of published literature is available in regard to the field aspects of source location theory and technique in the seismological and geotechnical areas. Here the analysis is for the most part three-dimensional in nature. Furthermore, since the structures involved are large and complex, there are a considerable number of assumptions and simplifications necessary. Under laboratory conditions one- and two-dimensional source location analyses are also of considerable importance. Unfortunately the literature in this area is relatively limited.

The prime purpose of this current chapter is to introduce the basic concepts of AE/MS source location, to briefly outline one-, two- and three-dimensional techniques, and to discuss the application of these techniques to laboratory and field studies in the geotechnical area. Limited consideration will also be given to the various computational methods (e.g., computer programs) associated with these techniques, however this aspect will not be considered in depth.

5.2   ONE-DIMENSIONAL SOURCE LOCATION

5.2.1   General

One-dimensional or linear source location is extensively utilized during AE/MS laboratory studies in the materials science field and to a lesser degree in the geotechnical area. The technique is mainly used where the test structures (specimens or models) are long compared to their transverse dimensions. For example, Figure 5.3 illustrates an experiment where it is desired to monitor the AE/MS activity during a uniaxial compression test and to determine the linear distribution of the associated AE/MS sources.

Figure 5.3a illustrates the experimental arrangement used for linear source location, which involves two AE/MS transducers attached near the upper and lower ends of the test specimen. As discussed later in this section the position of these transducers define the extent of the monitored area. The outputs of the transducers are connected to an AE/MS monitoring system containing a suitable parametric development module, the output of which provides the required linear source location data. Figure 5.3b illustrates typical results from such a test, namely the linear distribution of the AE/MS source locations monitored during the test period.

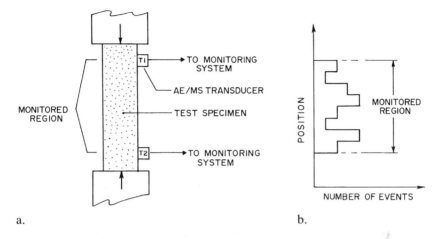

Figure 5.3. Experimental arrangement and typical data associated with the linear source location technique. a. Experimental arrangement for linear source location. b. Linear distribution of AE/MS sources.

5.2.2 One-dimensional location theory

a. *Conventional transducer placement*: The necessary transducer array required for one-dimensional source location is shown in Figure 5.4. Here the two transducers are located at x_1 and x_2, and a source is assumed to be located at x_s.

Now in the one-dimensional case the general three-dimensional source location equation (Eq. 5.4) reduces to the following,

$$V_i t'_i = \pm (x_i - x_s)$$ (Eq. 5.7)

and since there are two transducers (i = 1,2) this represents two equations

$$V_1 t'_1 = \pm (x_1 - x_s) = \mp (x_s - x_1)$$ (Eq. 5.8)

and

$$V_2 t'_2 = \pm (x_2 - x_s)$$ (Eq. 5.9)

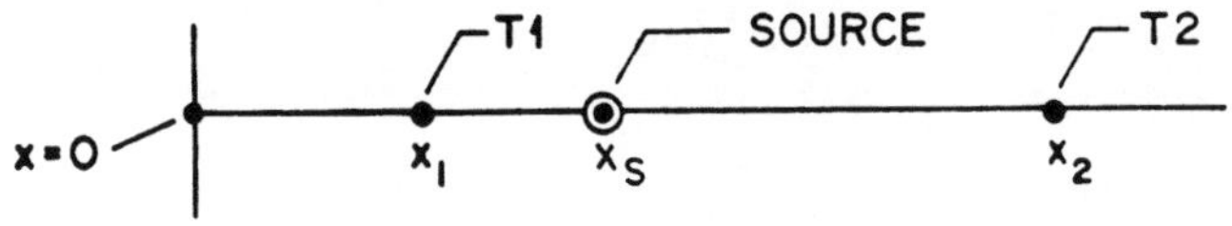

Figure 5.4. Relative positions of transducers and AE/MS source.

now from Equation 5.8, since $V_1 t'_1 > 0$ and $x_s > x_1$

$$V_1 t'_1 = + (x_s - x_1) \qquad \text{(Eq. 5.10)}$$

and from Equation 5.9, since $V_1 t'_2 > 0$, and $x_2 > x_o$

$$V_2 t'_2 = + (x_2 - x_s) \qquad \text{(Eq. 5.11)}$$

Now subtracting Equation 5.11 from 5.10 results in

$$V_1 t'_1 - V_2 t'_2 = x_s - x_1 - x_2 + x_s = 2x_s - (x_1 + x_2)$$

and setting $V_1 = V_2 = V$ we obtain

$$V (t'_1 - t'_2) = 2x_s - (x_1 + x_2) \qquad \text{(Eq. 5.12)}$$

Now let Δt denote the difference in arrival time at transducer T1 and T2, namely:

$$\Delta t = t_1 - t_2 \qquad \text{(Eq. 5.13)}$$

and noting from Equation 5.5 that

$$t'_i = t_i - T_s$$

or

$$t_i = t'_i + T_s$$

then Equation 5.13 becomes

$$\Delta t = t_1 - t_2 = (t'_1 + T_s) - (t'_2 + T_s)$$

$$\Delta t = t'_1 - t'_2 \qquad \text{(Eq. 5.14)}$$

substituting Equation 5.14 into 5.12 we obtain

$$V \Delta t = 2x_s - (x_1 + x_2)$$

or

$$x_s = \frac{1}{2} V \Delta t + \frac{(x_1 + x_2)}{2} \qquad \text{(Eq. 5.15)}$$

Equation 5.15 indicates that the position of the source (x_s) may be determined by measuring the time difference (Δt) of AE/MS events arriving at the transducers located at x_1 and x_2. It is important to note that the velocity (V) is necessary for source location, and that Equation 5.15 is only valid when the source lies within the array (i.e., $x_1 \leq x_s \leq x_2$).

It is interesting to consider a number of special situations in regard to Equation 5.15.

Case 1. $\Delta t = 0$.

Equation 5.15 with $\Delta t = 0$ reduces to

$$x_s = \frac{(x_1 + x_2)}{2}$$

indicating that the event occurred at a point half-way between the two transducers.

Case 2. $x_o = x_1$ (event at transducer T1).

From Equation 5.15 with $x_s = x_1$

$$x_1 = \frac{1}{2} V\Delta t + \frac{(x_1 + x_2)}{2}$$

$$\Delta t = \frac{(x_1 - x_2)}{V} = -\frac{(x_2 - x_1)}{V} \qquad \text{(Eq. 5.16)}$$

Case 3. $x_s = x_2$ (event at transducer T2)

Here Equation 5.15 with $x_s = x_2$ reduces to

$$\Delta t = +\frac{(x_2 - x_1)}{V} \qquad \text{(Eq. 5.17)}$$

Note that, although the results of Cases 2 and 3 are numerically the same, the signs are different. Generally the results obtained for Cases 1, 2, and 3 appear logical in terms of the location concept utilized.

It is also interesting to consider the special cases illustrated in Figure 5.5 where the source (x_s) is located outside the region between the two transducers. However for such cases (i.e., $x_s < x_1$ or $x_s > x_2$) Equation 5.15 is no longer valid and we must therefore return to the basic Equations 5.8 and 5.9.

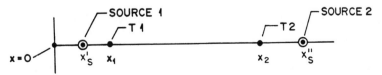

Figure 5.5. Relative positions of transducers and AE/MS sources for situation where sources are not located between the transducers.

For Equation 5.8

$$V_1 t'_1 = \pm (x_1 - x_s)$$

and since $V_1 t'_1 > 0$ and $x'_s < x_1$

$$V_1 t'_1 = + (x_1 - x'_s) \qquad \text{(Eq. 5.18)}$$

For Equation 5.9

$$V_2 t'_2 = \pm (x_2 - x'_s)$$

and since $V_2 t'_2 > 0$, and $x'_s < x_2$

$$V_2 t'_2 = + (x_2 - x'_s) \qquad \text{(Eq. 5.19)}$$

Now subtracting Equation 5.19 from 5.18 we obtain

$$V_1 t'_1 - V_2 t'_2 = (x_1 - x'_s) - (x_2 - x'_s)$$

and assuming $V_1 = V_2 = V$, and setting $\Delta t = t'_1 - t'_2$ from Equation 5.14 we obtain

$$V\Delta t = x_1 - x_2 = - (x_2 - x_1)$$

and rearranging we obtain

$$\Delta t = - \frac{(x_2 - x_1)}{V} \qquad \text{(Eq. 5.20)}$$

Similarly it may be shown that for the situation where $x''_s > x_2$

$$\Delta t = + \frac{(x_2 - x_1)}{V} \qquad \text{(Eq. 5.21)}$$

The interesting feature of Equations 5.20 and 5.21 is that the values of the actual source locations (x'_s and x''_s) do not appear, and the values of Δt are constant depending only on the velocity (V) and the transducer spacing ($x_2 - x_1$).

The preceding analysis indicates the following:
1. If the event occurs outside the array (i.e., if $x_s < x_1$ or $x_s > x_2$) a constant value of Δt will be obtained independent of the location. Therefore it is not possible to locate an event outside of the region between T1 and T2.

2. Based on the definition of Δt (i.e., $\Delta t = t_1 - t_2$) the sign of Δt (+ or -) will indicate the general region of the event, namely $\Delta t < 0$, $x_s < x_1$ and $\Delta t > 0$, $x_s > x_2$.

3. The result, however, provides an excellent method for computing velocity (V) for the material under study using a suitable mechanical calibration signal. Rearranging Equations 5.20 and 5.21 we obtain

$$V = \left| \frac{x_2 - x_1}{\Delta t} \right| \qquad \text{(Eq. 5.22)}$$

and the computed velocity will be independent of where the associated mechanical pulse is applied, provided $x_s < x_1$ or $x_s > x_2$.

Based on the results of the preceding analysis it is apparent from Figure 5.5 that it would be impossible to use a linear source location technique to locate AE/MS events occurring near the ends of a test specimen using transducers attached directly to the specimen.

b. *Remote transducer placement*: As indicated earlier in section (a), the conventional method of one-dimensional (linear) source location is unsatisfactory for locating AE/MS sources lying outside of the transducer array. This is a serious limitation when tests are carried out on specimens of geologic materials, since considerable undesirable background AE/MS activity, due to the rock/metal interface and a prominent non-uniform stress zone often occurs at or near the ends of such specimens. As a result source location in this region is important for the development of techniques for the reduction or elimination of background activity, and therefore a method for locating sources occurring in this region is highly desirable.

Figure 5.6a shows the monitored region associated with conventional transducer placement. Such transducer location results in two neglected regions where source location data cannot be obtained. The remote transducer mounting arrangement, illustrated in Figure 5.6b, appears to provide a means of overcoming the limitations of the conventional method.

Figure 5.7 illustrates the position of the various loading platen/specimen interfaces, (x_A and x_B), transducer locations (x_1 and x_2), and the location of a typical source (x_s). The appropriate stress wave velocities for the specimen (C_2) and the loading platens (C_1) are also indicated. The present analysis will only consider the case where the source is located between the transducers (i.e., $x_1 \leq x_s \leq x_2$).

The analysis follows that presented for the conventional arrangement, utilizing Equation 5.7, namely

$$V_i t'_i = \pm (x_s - x_i)$$

However since the associated stress waves traverse materials with different velocities (e.g., C_1 and C_2) the equation must be modified to the following

$$C_1 t'_{i1} + C_2 t'_{i2} = \pm (x_s - x_i) \qquad \text{(Eq. 5.23)}$$

where t'_{i1} and t'_{i2} denote the time the stress wave traverses the loading platens and the specimen, respectively, and noting the following

$$t'_i = t'_{i1} + t'_{i2} = \qquad \text{(Eq. 5.24)}$$

Now setting i = 1, 2 in Equation 5.23 we obtain

$$C_1 t'_{11} + C_2 t'_{12} = \pm (x_s - x_1) \qquad \text{(Eq. 5.25)}$$

and

$$C_1 t'_{21} + C_2 t'_{22} = \pm (x_s - x_2) = \mp (x_2 - x_s) \qquad \text{(Eq. 5.26)}$$

since $C_1 t'_{11} + C_2 t'_{12} > 0$ and $C_1 t'_{21} + C_2 t'_{22} > 0$ Equations 5.25 and 5.26 become

$$C_1 t'_{11} + C_2 t'_{12} = + (x_s - x_1) \qquad \text{(Eq. 5.27)}$$

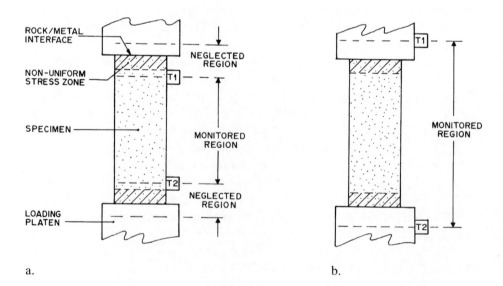

Figure 5.6. Conventional and remote transducer placement for one-dimensional source location monitoring. a. Conventional transducer location. b. Remote transducer location.

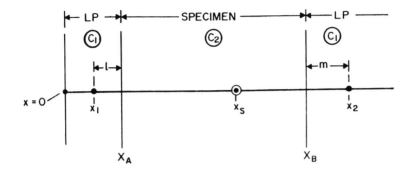

Figure 5.7. Relative positions of platen/specimen interfaces, transducers, and AE/MS source associated with remote transducer source location). (LP denotes loading platens, other symbols are defined in the text).

and

$$C_1 t'_{21} + C_2 t'_{22} = + (x_2 - x_s) \qquad \text{(Eq. 5.28)}$$

Subtracting Equation 5.28 from 5.27 we obtain

$$C_1 t'_{11} + C_2 t'_{12} - C_1 t'_{21} - C_2 t'_{22} = 2x_s (x_1 + x_2) \qquad \text{(Eq. 5.29)}$$

now from Equation 5.5

$$t'_i = t_i - T_s$$

and from Equation 5.24

$$t'_i = t'_{i1} - t'_{i2}$$

Therefore

$$t'_{i1} + t'_{i2} = t_i - T_o \qquad \text{(Eq. 5.30)}$$

Therefore for the two transducers we may write

$$t'_{11} + t'_{12} = t_1 - T_s$$

or

$$t'_{12} = t_1 - T_s - t'_{11} \qquad \text{(Eq. 5.31)}$$

and

$$t'_{21} + t'_{22} = t_2 - T_s$$

or

$$t'_{22} = t_2 - T_s - t'_{21} \qquad \text{(Eq. 5.32)}$$

and substituting Equations 5.31 and 5.32 into Equation 5.29 we obtain

$$C_1 t'_{11} + C_2 (t_1 - T_s - t'_{11}) - C_1 t'_{21} - C_2 (t_2 - T_s - t'_{21})$$

$$= 2x_s - (x_1 + x_2)$$

and rearranging we obtain

$$C_1(t'_{11} - t'_{21}) + C_2 [(t_1 - t_2) - (t'_{11} - t'_{21})]$$

$$= 2x_s - (x_1 + x_2)$$

and setting $t_1 - t_2 = \Delta t$ and rearranging further we obtain

$$x_s = \frac{1}{2} [C_2\Delta t + (x_1 + x_2) + (C_1 - C_2)(t'_{11} - t'_{21})] \qquad \text{(Eq. 5.33)}$$

Now the terms t'_{11} and t'_{21} represent the time the stress wave travels in the loading platens and this time is fixed depending only on C_1 and ℓ and m, namely

$$t'_{11} = \frac{x_A - x_1}{C_1} = \frac{\ell}{C_1} \qquad \text{(Eq. 5.34)}$$

and

$$t'_{21} = \frac{x_2 - x_B}{C_2} = \frac{m}{C_1} \qquad \text{(Eq. 5.35)}$$

Now substituting Equations 5.34 and 5.35 into Equation 5.33 and rearranging we obtain

$$x_s = \frac{1}{2} \left[C_2\Delta t + (x_1 + x_2) + (C_1 - C_2)\left(\frac{\ell - m}{C_1}\right) \right]$$

or

$$x_s = \frac{1}{2} C_2 \Delta t + \frac{(x_1 + x_2)}{2} + \frac{\ell - m}{2} \left(1 - \frac{C_2}{C_1}\right) \qquad \text{(Eq. 5.36)}$$

Now considering a number of special cases, namely:

Case 1. $C_1 = C_2$.

$$x_s = \frac{1}{2} C_2 \Delta t + \left(\frac{(x_1 + x_2)}{2}\right)$$

where the result is the same as for the conventional transducer arrangement (Eq. 5.15).

Case 2. $\ell = m$.

$$x_s = \frac{1}{2} C_2 \Delta t + \left(\frac{(x_1 + x_2)}{2}\right)$$

With transducers located at equal distances on steel loading platens, the solution reduces to that of conventional arrangement (Eq. 5.15).

Case 3. $C_1 \rightarrow \infty$, $\ell \neq m$.

$$x_s = \frac{1}{2} C_2 \Delta t + \left(\frac{(x_1 + x_2)}{2}\right) + \left(\frac{\ell - m}{2}\right)$$

which reduces to

$$x_s = \frac{1}{2} C_2 \Delta t + \left(\frac{x_A + x_B}{2}\right)$$

Therefore, when $C_1 \rightarrow \infty$ the effective transducer locations move to x_A and x_B.

Case 4. $C_1 \rightarrow 0$, $\ell \neq m$.

$$x_s = \frac{1}{2} C_2 \Delta t + \frac{(x_1 + x_2)}{2} + \left(\frac{\ell - m}{2}\right)\left(1 - \frac{C_2}{C_1}\right)$$

Now for $C_1 \rightarrow 0$, $x_s \rightarrow \infty$ and solution has no meaning since the stress wave never reaches the transducers.

c. *Experimental determination of* Δt: Figure 5.8 illustrates an experimental technique for determination of the value of Δt required for source location calculation using the one-dimensional (linear) technique. Figure 5.8a illustrates the conventional transducer placement method (using Eq. 5.15), however the technique is also applicable to remote transducer placement (using Eq. 5.36).

In Figure 5.8b suitable conditioned analog signals from the two transducers attached to the specimen (T1 and T2) enter the arrival-time-difference module, pass through their respective threshold detector circuits, and generate output pulses at times t_1 and t_2. For Δt measurements it is necessary that precautions are taken (e.g. suitably high threshold level or the addition of accessory electronics) to insure that event pulses, rather than ring-down pulses, are generated. The time between the pulses at t_1 and t_2 is equal to the required Δt. The first pulse (either t_1 or t_2) triggers the gate circuit open and the counter begins accumulating clock pulses. The second pulse (either t_2 or t_1) closes the gate circuit leaving the counter with a total of N pulses accumulated during the time Δt.

The function of the logic circuit shown in Figure 5.8b is three-fold. First it must route the first detected pulse (t_1 or t_2 depending on the location of the source) to the "start input" of the gate, and the second detected pulse to the "stop input". Second it must generate a signal (+ or -) to indicate whether t_1 or t_2 was detected first in order to determine the polarity of Δt (since $\Delta t = t_1 - t_2$ if

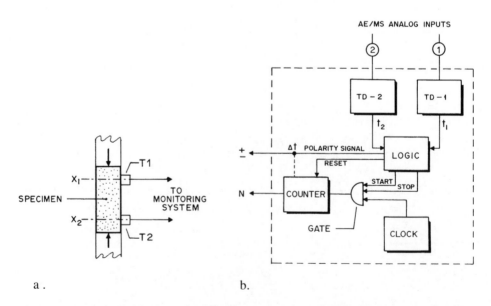

a.   b.

Figure 5.8. Experimental determination of arrival-time-difference (Δt)data for one-dimensional source location. a. Experimental arrangement. b. Arrival time difference (Δt) module.

$t_1 > t_2$, $\Delta t > 0$ and if $t_1 < t_2$, $\Delta t < 0$). Thirdly the logic circuit, prior to receiving the signals associated with the next event, must reset the counter to zero.

The output of the arrival-time-difference (Δt) module is the total number of pulses (N) accumulated during the time Δt. The value of Δt may then be computed using the expression

$$\Delta t = \pm N/R \qquad \text{(Eq. 5.37)}$$

where R is the clock rate in pulses per second (pps). For example if N = 375 and R = 10^6 pps then

$$\Delta t = \pm 375/10^{-6} = \pm 375/10^{-6}\text{s} = \pm 375\mu s$$

The polarity (+ or –) of Δt may be provided as a separate output signal and applied in subsequent circuitry for computation of Δt. In some systems the polarity signal may be applied directly to the counter to provide an output which includes the time difference polarity.

5.3   TWO-DIMENSIONAL SOURCE LOCATION

5.3.1   General

Two-dimensional or planar source location of AE/MS events is not utilized extensively in the geotechnical area. A number of applications do exist in the laboratory including studies on thin beams and plates and other cases where the specimen or model is small in one of its dimensions (e.g. diametric loading of thin discs during Brazilian tests). In the field, applications include the location of events occurring in a defined layer (e.g. coal or ore seam), or in surface or near-surface planar structures.

In contrast, a great majority of the non-geologic structures investigated using AE/MS technique are effectively two-dimensional in form (e.g. pressure vessels, pipelines, metal and concrete structures and structural elements, etc.). For such studies a variety of planar source location techniques have been developed, and automated systems for source location analysis are available from a number of AE/MS equipment manufacturers.

5.3.2   Basic concepts

Figure 5.9 shows two transducers (T1 and T2) located on a planar structure, and an AE/MS source located at the point $P_s(x_s, y_s)$. A simple analysis of this situation provides a useful illustration of the basic concepts involved in planar location.

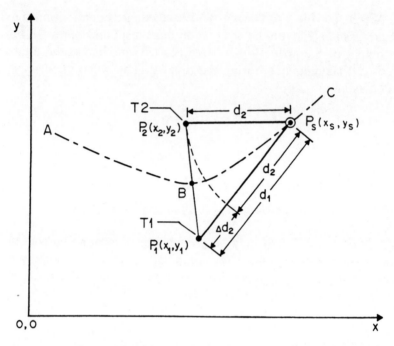

Figure 5.9. Geometry illustrating the basic concept of two-dimensional source location.

From Figure 5.9 it is clear that

$$d_1 = t'_1 V = (t_1 - T_s)V \qquad \text{(Eq. 5.38)}$$

and

$$d_2 = t'_2 V = (t_2 - T_s)V \qquad \text{(Eq. 5.39)}$$

where T_s is the source time, V is the stress wave velocity, d_1 and d_2 are the distances from the source to transducers T1 and T2, and t'_1 and t'_2 and t_1 and t_2 are the associated travel and arrival times, respectively. As illustrated in Figure 5.9 we may write

$$\Delta d = d_1 - d_2$$

and substituting from Equations 5.38 and 5.39 yields

$$\Delta d = (t_1 - T_s)V - (t_2 - T_s)V$$

Therefore

$$\Delta d = (t_1 - t_2)V = V\Delta t \qquad \text{(Eq. 5.40)}$$

According to Equation 5.40, in the case where the source lies at a point in a plane, the difference traveled by the stress wave to a pair of transducers may be calculated from the measured time difference. It is important to note that in a plane there are a number of points for which Equation 5.40 is valid, and the locus of these points defines the curve ABC shown in Figure 5.9. Such a curve is a hyperbola, which is defined as "the locus of a point which moves so that the difference of its distances from two field points (foci) is equal to a constant." In Figure 5.9 the transducers are located at the foci (i.e., P_1 and P_2), and a whole family of hyperbolas are associated with this set of foci depending on the value of $V\Delta t$ in Equation 5.40.

In planar source location two pairs of transducers (one transducer may be common to both pairs) are utilized to generate two hyperbolas, the intersection of which is the required source location. This situation is illustrated in Figure 5.10 where the source is located at the intersection of the two hyperbolas associated with transducer pairs T_1/T_2 and T_2/T_3.

It is important to note that ambiguous source locations may occur when the minimum number of transducers ($n = 3$) are used. Such ambiguity occurs in regions close to and "behind" each transducer where hyperbolas associated with each of the transducer pairs intersect twice yielding twin solutions. Figure 5.11a illustrates such a situation. Here the hyperbola $\Delta d_{12} = k_{12}$, associated

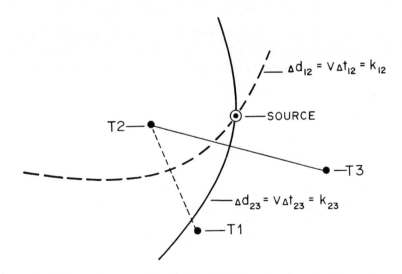

Figure 5.10. Source location determined by the intersection of two hyperbolas.

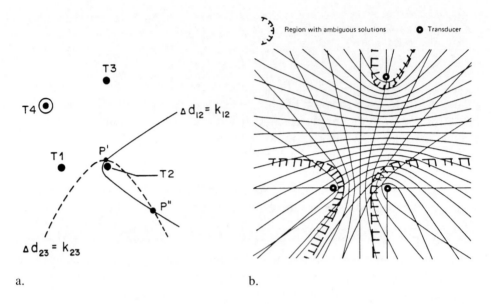

Figure 5.11. Ambiguous source locations occurring using a three transducer array. a. Hyperbola intersection generating two possible source locations (P' and P''). b. Ambiguous source location regions associated with a three transducer array (after Rindorf, 1981).

with transducer pair T1/T2, intersects the hyperbola $\Delta d_{23} = k_{23}$, associated with transducer pair T2/T3, at two points P' and P''. Figure 5.11b illustrates the regions of ambiguity associated with a typical three transducer source location array. In order to resolve the ambiguity (i.e. to select the correct source location) additional information is required. This can be obtained by measuring the delay time associated with an additional transducer (T4) and comparing it with the calculated value for the two possible source locations.

5.3.3 Source location algorithms

Rindorf (1981), in a comprehensive technical review of a number of aspects of AE/MS, discusses two-dimensional source location in considerable detail. He indicates that the algorithm used to calculate source locations from experimental data may be based on several methods, including iterative methods, look-up tables or a direct mathematical solution. He indicates that exact mathematical algorithms for a plane and spherical surface have been presented earlier by Tobias (1976) and Asty (1978). He also includes in his review a source location algorithm and associated program, developed for the HP 41-C calculator, for use with a three-transducer rectangular array.

a. *Algorithm development*: Here an algorithm will be developed on the basis of Equations 5.2 and 5.3 presented earlier. This approach has been discussed in some detail in papers by Miller and Harding (1972) and Blake et al. (1974). Figure 5.12 illustrates the associated position of the source, P_s (x_s, y_s), a transducer located at the point P_o (x_o, y_o), and a number of additional transducers located at the points P_i (x_i, y_i). The transducer at P_o (x_o, y_o) is assumed to be the closest transducer to the event source.

In general it is clear that

$$d_i = V_i (t_i - t_o) \qquad \text{(Eq. 5.41)}$$

where V_i is the stress wave velocity, t_i is the arrival time at the ith transducer, t_o is the arrival time at the closest transducer (located at P_o). Furthermore it is assumed that $t_o \leq t_i$ for i = 1, 2, 3. The starting point for the analysis is the expression

$$d_i + d_o = d'_i \qquad \text{(Eq. 5.42)}$$

Now substituting for d_o and d'_i in terms of the coordinates of the associated points Equation 5.42 becomes

$$d_i + [(x_s - x_o)^2 + (y_s - y_o)^2]^{\frac{1}{2}} = [(x_s - x_i)^2 + (y_s - y_i)^2]^{\frac{1}{2}} \quad \text{(Eq. 5.43)}$$

Now squaring Equation 5.43 and rearranging we obtain

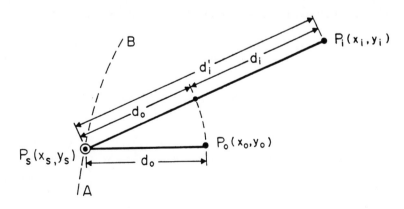

Figure 5.12. Geometry used for the development of the two-dimensional source location algorithm.

$$d_i^2 + [(x_s - x_o)^2 + (y_s - y_o)^2] - [(x_s - x_i)^2 + (y_s - y_i)^2]$$

$$= -2d_i [(x_s - x_o)^2 + (y_s - y_o)^2]^{\frac{1}{2}}$$

Expanding squared terms we obtain

$$d_i^2 + x_s^2 - 2x_s x_o + x_o^2 + y_s^2 - 2y_s y_o + y_o^2 - x_s^2 + 2x_s x_i$$

$$-x_i^2 - y_s^2 + 2y_s y_i - y_i^2 = -2d_i [(x_s - x_o)^2 + (y_s - y_o)^2]^{\frac{1}{2}}$$

Simplifying, this reduces to

$$(d_i - x_o^2 + y_o^2 - x_i^2 - y_i^2) - 2[(x_o - x_i)x_s + (y_o - y_i)y_s]$$

$$= -2d_i [(x_s - x_o)^2 + (y_s - y_o)^2]^{\frac{1}{2}}$$

Setting

$$K_i = d_i^2 + x_o^2 + y_o^2 - x_i^2 - y_i^2 \qquad \text{(Eq. 5.45)}$$

$$\alpha_i = x_o - x_i \qquad \text{(Eq. 5.46)}$$

$$\beta_i = y_o - y_i \qquad \text{(Eq. 5.47)}$$

we obtain

$$K_i - 2(\alpha_i x_s + \beta_i y_s) = -2d_i [(x_s - x_o)^2 - (y_s - y_o)^2]^{\frac{1}{2}} \quad \text{(Eq. 5.47)}$$

Assuming $d_i \neq 0$, and dividing both sides by $-2d_i$ we obtain

$$-\frac{K_i}{2d_i} + \frac{(\alpha_i x_s + \beta_i y_s)}{d_i} = [(x_s - x_o)^2 + (y_s - y_o)^2]^{\frac{1}{2}} \quad \text{(Eq. 5.48)}$$

For i = 1 in Equation 5.48 we obtain

$$-\frac{K_i}{2d_1} + \frac{(\alpha_1 x_s + \beta_1 y_s)}{d_1} = [(x_s - x_o)^2 + (y_s - y_o)^2]^{\frac{1}{2}} \quad \text{(Eq. 5.49)}$$

Now subtracting Equation 5.48 for i = 2,3 from Equation 5.49 we obtain

$$-\frac{K_i}{2d_1} + \frac{K_i}{2d_1} + \left(\frac{\alpha_1 x_s + \beta_1 y_s}{d_1}\right) - \left(\frac{\alpha_i x_s + \beta_i y_s}{d_1}\right) = 0$$

and rearranging we obtain

$$\frac{\alpha_1 x_s}{d_1} - \frac{\alpha_i x_s}{d_i} + \frac{\beta_1 y_s}{d_1} - \frac{\beta_i y_s}{d_i} = \frac{1}{2}\left(\frac{K_1}{d_1} - \frac{K_i}{d_i}\right)$$

and regrouping terms, we obtain

$$2\left(\frac{\alpha_1}{d_1} - \frac{\alpha_i}{d_i}\right)x_s + 2\left(\frac{\beta_1}{d_1} - \frac{\beta_i}{d_i}\right)y_s = \left(\frac{K_1}{d_1} - \frac{K_i}{d_i}\right) \quad \text{for i = 2,3}$$

Setting

$$A_i = 2\left(\frac{\alpha_1}{d_1} - \frac{\alpha_i}{d_i}\right) \qquad\qquad \text{(Eq. 5.50)}$$

$$B_i = 2\left(\frac{\beta_1}{d_1} - \frac{\beta_i}{d_i}\right) \qquad\qquad \text{(Eq. 5.51)}$$

$$H_i = \left(\frac{K_1}{d_1} - \frac{K_i}{d_i}\right) \qquad\qquad \text{(Eq. 5.52)}$$

The final result becomes

$$A_i x_s + B_i y_s = H_i \quad \text{For i = 2,3} \qquad\qquad \text{(Eq. 5.53)}$$

Now setting i = 2 and 3 provides two equations in two unknowns (x_s, y_s), the coordinates of the AE/MS source. These two equations can be solved by various methods such as Cramer's rule.

It should be noted that the factors a_i and b_i (Equations 5.46 and 5.47) depend only on the coordinates of the four transducers. However, K_i, A_i, B_i and H_i depend on the distances d_i (i = 1, 2, 3) which are a function of the arrival times t_i (i = 1, 2, 3) through Equation 5.41, namely

$$d_i = V_i (t_i - t_o)$$

Note that d_o does not appear in the final expressions.

It should also be noted that Equations 5.50, 5.51, and 5.52 involve terms containing $1/d_i$ (i = 1, 2, 3), and those cases were $d_i = 0$, A_i, B_i and $H_i \rightarrow \infty$. As a result solutions where the source is located close to a transducer may be seriously in error.

b. *Algorithm application*: In order to better appreciate the method by which the algorithm developed in the previous section is utilized, the various steps involved in calculation of the source coordinates will be outlined here. The necessary input data from an AE/MS experiment will be the transducer coordinates, (x_0, y_0), (x_1, y_1), $x_2, y_2)$ and (x_3, y_3); the stress wave propagation velocity, V_i, assumed here to be isotropic, and therefore $V_i = V$; and the experimental data from which the arrival time data may be computed. Figure 5.13 illustrates typical AE/MS signals, associated with a specific event, detected by a four-transducer array. The four-associated arrival times are seen to be t_0, t_1, t_2 and t_3.

The various steps involved in computing the coordinates of the AE/MS source generating the data shown in Figure 5.13 are as follows:

Step 1. From Equation 5.41, assuming $V_i = V$.

$$d_i = V(t_i - t_0) = V\Delta t_{oi}$$

and based on the preceding equation and the values of Δt_{oi} indicated in Figure 5.13 we may compute

$$d_1 = V\Delta t_{01}$$

$$d_2 = V\Delta t_{02}$$

$$d_3 = V\Delta t_{03}$$

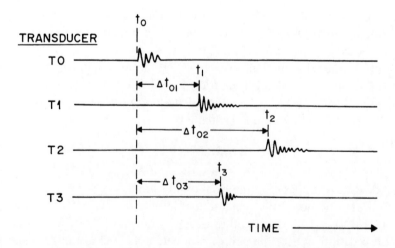

Figure 5.13. Arrival time data associated with a four-transducer array.

Step 2. Now since the coordinates of the various transducers [i.e. (x_0, y_0), (x_1, y_1), (x_2, y_2), (x_3, y_3)] are known along with the distance d_1, d_2, d_3 we may utilize Equations 5.45, 5.46, and 5.47, namely:

$$K_i = d_i^2 + x_0^2 + y_0^2 - x_i^2 - y_i^2$$

$$\alpha_i = x_0 - x_i$$

$$\beta_i = y_0 - y_i$$

to compare the factors K_1, K_2, K_3; α_1, α_2, α_3; and β_1, β_2, β_3.

Step 3. Now setting the calculated values of K_i, a_i and b_i from step 2 into Equations 5.50, 5.51 and 5.52, namely:

$$A_i = 2 \left(\frac{\alpha_1}{d_1} - \frac{\alpha_i}{d_i} \right)$$

$$B_i = 2 \left(\frac{\beta_1}{d_1} - \frac{\beta_i}{d_i} \right)$$

$$H_i = \left(\frac{K_1}{d_1} - \frac{K_i}{d_i} \right)$$

where $i = 2, 3$ in all cases, we may calculate the factors A_2, A_3, B_2, B_3; and H_2, H_3.

Step 4. Substituting the calculated values of A_i, B_i and H_i ($i = 2, 3$) into Equation 5.53 we obtain two equations, namely:

$$A_2 x_s + B_2 y_s = H_2$$

$$A_3 x_s + B_3 y_s = H_3$$

Step 5. Finally the two previous equations may be solved by Cramer's method, to yield the source location coordinates, namely:

$$x_s = \frac{\begin{vmatrix} H_2 & B_2 \\ H_3 & B_3 \end{vmatrix}}{\begin{vmatrix} A_2 & B_2 \\ A_3 & B_3 \end{vmatrix}} = \frac{A_2 B_3 - H_3 B_2}{A_2 B_3 - A_3 B_2}$$

and
$$y_s = \frac{\begin{vmatrix} A_2 & H_2 \\ A_3 & H_3 \end{vmatrix}}{\begin{vmatrix} A_2 & B_2 \\ A_3 & B_3 \end{vmatrix}} = \frac{A_2 H_3 - A_3 H_2}{A_2 B_3 - A_3 B_2}$$

5.4   THREE-DIMENSIONAL SOURCE LOCATION

5.4.1   General

Although true three-dimensional source location techniques are rarely used in the material science or non-destructive testing areas, these techniques are utilized in the majority of AE/MS studies carried out in the seismological and geotechnical fields. As a result, for the most part, the literature relative to this technique is found in the areas of seismology, mining engineering and rock mechanics. During the last 15 years there has been a considerable number of studies carried out in order to improve source location analysis techniques. Historically, it is interesting to note that in his early paper on rock burst location in South Africa, Cook (1963), although familiar with the analytical source location techniques utilized in seismology, made use of a mechanical system for source location based on the classic "string model". Sometime later suitable computer-based source location techniques were developed in South Africa and an excellent description of one such technique is given in the paper by Salamon and Wiebols (1974).

A paper by Blake and Leighton (1970) describes the early source location techniques utilized at the USBM for use in underground mines. Later papers by Leighton and Duvall (1972) and Blake et al. (1974) discuss the subject in further detail and present a number of improvements in the technique. A more recent USBM publication by Dechman and Sun (1977) describe the development of source location techniques for application to surface problems such as slope stability evaluation in open pit mines.

A paper by Kijko (1975) has been concerned with the development of source location programs and the optimization of transducer arrays to provide maximum source location accuracy. This paper provides an excellent review of source location analysis techniques and includes an analytical method for designing optimal transducer arrays. Although the studies was undertaken in relation to regional seismic studies, it is directly applicable to a variety of geotechnical studies. Kijko in fact applies a number of the techniques to source location studies associated with the Szombierki coal mine in Poland, as well as to regional seismic studies in the Lublin, Upper Silesia, and Ostrava coal basins.

A number of other papers concerned with various aspects of AE/MS source location analysis are of interest. For example, the paper by Miller and Harding

(1972) presents an analysis of the possible errors involved in source location; and a later paper by Cete (1977) is concerned with the mathematical aspects of source location and the effects of errors in the velocity and arrival time data. About the same time at Penn State, an in-source location program (IBMSL) was developed (Mowrey, 1977) and was programmed for use on the Penn State IBM 370/3033 digital computer. This program is based on the assumption that the AE/MS source and the monitoring transducers are located in a continuous, homogenous medium, although the program is capable of including anisotropic velocity characteristics if necessary. In order to carry out source location calculations in bedded strata, a second program (HYPO71) was obtained from the U.S. Geological Survey (USGS) and modified for use on the Penn State IBM 370/3033 computer (Hardy et al., 1981).

5.4.2 Source location algorithms

a. *Algorithm development*: As illustrated in Figure 5.14 the three-dimensional source location concept is simply an extension of that used in two dimensions. Blake et al. (1974) describes the development of the required algorithm, involving the use of a five transducer array, which follows the same general approach used earlier for two-dimensions (see Section 5.3.3). The results of the analysis, based on p-wave arrival data and presented in a format similar to that used for the two-dimensional case, reduces to the solution of three equations in three unknowns (i.e., the source coordinates), namely:

$$A_i x_s + B_i y_s + C_i Z_s = H_i \quad \text{(For i = 2, 3, 4)} \quad \text{(Eq. 5.54)}$$

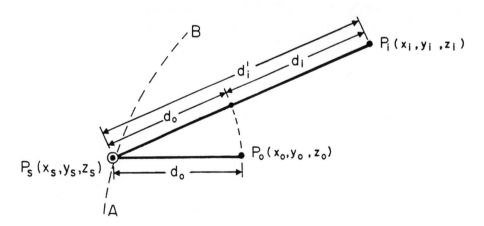

Figure 5.14. Geometry used for the development of the three-dimensional source location algorithm.

where

$$A_i = 2 \left(\frac{\alpha_1}{d_1} - \frac{\alpha_i}{d_i} \right) \qquad \text{(Eq. 5.55)}$$

$$B_i = 2 \left(\frac{\beta_1}{d_1} - \frac{\beta_i}{d_i} \right) \qquad \text{(Eq. 5.56)}$$

$$C_i = 2 \left(\frac{\gamma_1}{d_1} - \frac{\gamma_i}{d_i} \right) \qquad \text{(Eq. 5.57)}$$

$$H_i = \left(\frac{K_1}{d_1} - \frac{K_i}{d_i} \right) \qquad \text{(Eq. 5.58)}$$

and

$$\alpha_i = x_o - x_i \qquad \text{(Eq. 5.59)}$$

$$\beta_i = y_o - y_i \qquad \text{(Eq. 5.60)}$$

$$\gamma_i = z_o - z_i \qquad \text{(Eq. 5.61)}$$

$$K_i = d_i^2 + x_o^2 + y_o^2 + z_o^2 - x_i^2 - y_i^2 - z_i^2 \qquad \text{(Eq. 5.62)}$$

where the various factors in Equations 5.55 to 5.62 are obtained by setting $i = 2, 3, 4$.

As in the two-dimensional case the value of d_o does not appear in the final solution, although the coordinates of the closest transducer (i.e. x_o, y_o, z_o) are involved. As in the two-dimensional case the arrival time differences $\Delta t_{oi} = (t_i - t_o)$ enter the solution through the calculation of the d_i values using the expression

$$d_i = V_i (t_i - t_o) = V_i \Delta t_{oi} \qquad (i = 1, 2, 3, 4) \qquad \text{(Eq. 5.63)}$$

It should be noted that those solutions, where the source is located near one of the transducers, may be seriously in error due to the presence of the $1/d_i$ terms in Equations 5.55 to 5.58.

b. *Algorithm application*: The algorithm described in the proceeding section is applied in the same manner as illustrated for the two-dimensional case earlier in Section 5.3.3. In the three-dimensional case it is necessary to solve the three simultaneous equations, represented by Equations 5.64 for $i = 2, 3, 4$, namely:

$$A_2 x_s + B_2 y_s + C_2 z_s = H_2$$

$$A_3 x_s + B_3 y_s + C_3 z_s = H_3 \qquad \text{(Eq. 5.64)}$$

$$A_4 x_s + B_4 y_s + C_4 z_s = H_4$$

This may be accomplished using Cramer's rule. The various steps involved in computing source coordinates is as follows:

Step 1. Calculate d_i values from Equation 5.63 using experimentally determined values of Δt_{oi}.

Step 2. Calculate α_i, β_i, γ_i and K_i values from Equations 5.59 to 5.62 using transducer coordinates and d_i values from Step 1.

Step 3. Calculate A_i, B_i, C_i and H_i values from Equations 5.55 to 5.58 using d_i values from Step 1 and α_i, β_i, γ_i and K_i values from Step 2.

Step 4. Substitute A_i, B_i, C_i and H_i values from Step 3 into Equations 5.64 and solve for x, y, z using Cramers rule.

5.5 APPLICATION OF LEAST SQUARES METHOD

The source location methods for one-, two- and three-dimensions discussed in the previous sections utilize data from a minium number of transducers, i.e. N=5 for three-dimensional location. Studies by Leighton and Duvall (1972) indicate that the use of additional transducers and least squares techniques provides more accurate and reliable results. This technique also appears to be particularly effective in improving solutions for AE/MS sources located outside the transducer array boundaries. Most present day location procedures utilize additional transducers and some form of statistical analysis.

Using model and field data Leighton and Duvall investigated the improvement in location accuracy (reduction of location error) using various numbers of transducers (geophones). Here the direct solution (5 geophones) and the least squares solution were compared by computing source locations using both methods. The input data was based on a model geophone array for which the computed arrival times at each transducer were utilized. Location accuracy for events occurring both inside and outside the array were computed. The factor nD in the Figure 5.15 describes the location of the test source relative to the array dimensions. D is defined as the distance between the highest and lowest transducer in the array. Figure 5.15 illustrates the variation in error for model studies, where in case 1 correct arrival data was used for all transducers and in Case 2, a 10% error was introduced in the arrival time data for one transducer.

In Figure 5.15a, results for 2D, 3D and 5D are shown. Source location error at the edge and inside the array were very small and basically the same for direct and least squares methods. This data is not included; clearly the error decreases rapidly as the number of transducers used to provide data increases. Here for example for 3D the addition of one transducer (5 to 6) decreases the location error by approximately 50%.

In Figure 5.15b results for sources located at the edge of the array, and for 1D, 3D, 5D, 7D and 9D are shown. In this case, where a 10% error was introduced in one of the transducer arrival times, the overall source location errors are seen to increase by factors of 4 or more over Case 1. However, as in Case 1, the use of an increasing number of transducers results in a rapid decrease in location error. For example for 3D the addition of one transducer (5 to 6) decreases the location error by a factor of approximately 1.8. Similar studies were carried out using actual field data. Although the errors were considerably higher, the general results were similar to those noted for the model data.

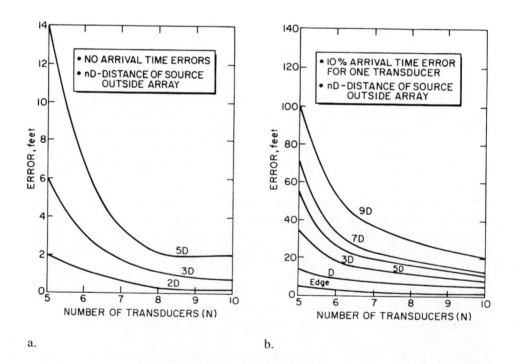

a.   b.

Figure 5.15. Variation in source location error, based on model studies, as a function of the distance between the source and the array, the number of transducers used in the calculation, and the effect of arrival time error in one transducer (after Leighton and Duvall, 1972). (Length conversion factor: 1 ft = 0.305 m.) a. Case 1 - no arrival time errors. b. Case 2 – 10% arrival time error for one transducer.

Based on the preceding studies the authors' drew the following general conclusions:

1. Direct solutions are improved both inside and outside the geophone array boundaries when the least square method is employed.
2. The least squares treatment of data outside the array boundaries greatly improves the reliability of solutions at any distance from these boundaries.
3. The least squares solutions become more exact as the number of geophones used in the solution increases.

Over the years there has been considerable research undertaken relative to the topic of source location. While Leighton and Duvall's original conclusions remain valid, the effects of other factors, in particular, transducer array geometry, have been found to be extremely important. This factor and others will be discussed later in the chapter.

5.6 EXPERIMENTAL DETERMINATION OF ARRIVAL-TIME-DIFFERENCE DATA

Automated systems have been developed to provide the multi-channel arrival-time-difference data (Δt_{oi}) required for three-dimensional source location. Such systems are based on the same general concept as that described earlier (Section 5.2.2) for use in one-dimensional source location. Figure 5.16 illustrates the format of the multi-channel system, termed a "rock burst monitor" (RBM), developed by the USBM (Blake, 1971) for use in geotechnical field studies.

Basically, the system consists of a set of event (threshold) detectors, a clock, counters, a coincidence detector, and a timing and control section. In operation, when any one of the event detectors receive an analog signal greater than a predetermined threshold level (±1.0V), all counters are activated and begin counting clock pulses at a rate of 10,000/s. As each subsequent event detector receives an input signal greater than the threshold level, the counter for that channel is stopped and the associated count held. After a selected time interval (< 0.99s) from initiation of the detection cycle, the counters are scanned to determine whether the event was detected on at least four channels. If so the counter data is printed out; if not, the counters are reset and are ready to process the next event (i.e. to begin another detection cycle). Minimum cycle time is 2s, allowing a maximum of 30 events per minute to be processed. If external electronic interference occurs, two or more event detectors will be triggered simultaneously. Such a situation is automatically sensed by the built-in coincidence detector, the false event is ignored and the system reset. The system also incorporates an additional counter circuit which provides time-of-day data for each valid event. Later versions of the RBM incorporated

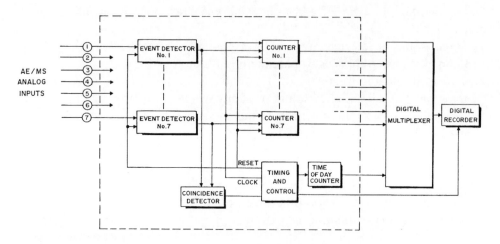

Figure 5.16. Simplified block diagram of the USBM "rock burst monitor-early model" (after Blake, 1971).

additional circuitry to allow the event energy to be estimated, using a signal integrating procedure, and result to be printed out.

Similar systems have been developed for laboratory use, and differ mainly in their event rate capability and timing resolution. For example the laboratory system described by Byerlee and Lockener (1977) is capable of processing of the order of 18,000 AE/MS events per minute.

5.7   USE OF P- AND S-WAVE DATA

The source location algorithms described in Sections 5.3 and 5.4 have been developed on the basis of p-wave arrival data.* Based on direct analysis procedures these require a minimum of four transducers for two-dimensional analysis, and five for three-dimensional analysis. In those situations where suitable s-wave as well as p-wave arrival data are available modified algorithms may be utilized. These will reduce the number of transducers required for two- and three-dimensional analysis to three and four, respectively. Blake et al. (1974) indicate how this may be accomplished. It should be pointed out, however, that this will require a knowledge of both the p- and s-wave velocities. Furthermore, where the transducer-to-source distances are small, the p-wave/s-wave separation may not be sufficient to utilize this approach.

---

* The analysis presented in Sections 5.3 and 5.4 may in fact utilize any suitable component of the stress wave (i.e. p-wave, s-wave, Rayleigh Wave, etc.), provided the data is of satisfactory quality and the associated velocity is available.

5.8   VELOCITY MODELS

A prime element in all source location techniques is the appropriate stress wave velocity. In the discussions presented earlier in this chapter it has generally been assumed that this velocity was isotropic** (i.e., independent of direction) and that it remained constant throughout the study. These assumptions, although valid in specific situations, are certainly not true in general, and in many cases, a "velocity model", rather than a single value of velocity may be required.

Past laboratory and field studies have clearly indicated that stress wave velocity is dependent upon many factors including rock type, porosity and density of the rock matrix, temperature, pressure and stress level (e.g., velocity increases with increasing depth), density of the fluid in the pores of the rock, and the dispersive properties of the rock itself (e.g., each frequency component in a stress wave travels at its own velocity). The need for a suitable velocity model to adequately describe the specific laboratory or field situation is therefore obvious.

For example, Rothman (1975) observed that during laboratory creep studies on a sandstone that there were substantial differences in the velocities in the x-, y- and z-directions, and that these velocities varied with specimen deformation. Based on this he concluded that in such studies accurate source location was only possible if these velocity characteristics (i.e., a velocity model) were incorporated in the source location analysis.

Similar situations arise in the field. For example in a recent study of AE/MS activity associated with a longwall coal mine (Hardy et al., 1978) it was apparent that at least three regions having different velocity characteristics were involved, namely:
1. The gob or fractured region behind the longwall face.
2. The immediate region around and above the longwall face.
3. The virgin region ahead of the longwall face.
Since such a longwall operation is a dynamic process, and as the longwall face advances the three velocity regions move with it. Since the transducer array itself remains fixed, it is clear that the rather complex three-dimensional velocity model associated with the mine structure is continually changing its characteristics as the coal face advances.

As described by Rothman (1975) the required velocity models may be determined in the laboratory by attaching small piezoelectric elements, in the form of transmitter-receiver pairs, to the test structure (i.e., specimen or model) and measuring the associated transit times. These times along with the associated distances provide a means of calculating the required velocity data. Such measurements carried out at various times throughout an experiment

---

** The basic algorithms, however, were developed to allow for anisotropic velocity through the distance-time relations (e.g. Equation 5.41, $d_i = V_i (t_i - t_o)$), which included a specific velocity V_i for each transducer.

allow the necessary velocity model to be developed and updated as the test proceeds.

In many AE/MS field studies, for example those associated with mine and tunnel stability, the required velocity model is normally obtained by underground detonation of small explosive charges at known locations. By monitoring the AE/MS signals generated in such studies, using the installed AE/MS transducer array, and carrying out the associated computations, it is possible to develop a suitable velocity model for the structure under study. This procedure is not practical in some field situations (e.g., gas storage reservoirs, solution mined caverns, etc.) and surface seismic and down-hole velocity logging techniques must be employed (see Hardy et al., 1981). A paper by Hardy (1986) discusses a range of techniques for development of suitable velocity models.

A review of the current literature indicates that considerable experimental and analytical research is required to improve the quality of velocity models utilized in AE/MS field studies. The unavailability of accurate and well-defined velocity models probably represents the most critical limiting factor in a number of AE/ MS field applications.

5.9   SOURCE LOCATION ERRORS

In general, errors in source location arise from a number of factors including the following:
1. Accuracy of the velocity model utilized.
2. Ability to time and read arrival times.
3. Transducer array geometry.
4. Array size.
5. Errors in transducer coordinates.
6. Location of the source relative to the array.

In the following paragraphs a number of these factors will be considered in more detail.

5.9.1   Velocity model

As discussed earlier in Section 5.8, a suitable velocity model is of prime importance in minimizing source location errors. Although a suitable model may be relatively easily obtained for studies under laboratory conditions, velocity models for many field situations may be highly complex, and even crude approximations may be difficult to obtain. The relationship of transducer array geometry and the quality of the associated velocity model will be considered further in Section 5.9.4.

5.9.2   Arrival time

The ability to accurately determine arrival times is governed by factors such as the signal-to-noise ratio, the frequency of the AE/MS signal waveform, and the associated timing system. Problems associated with the above factors are often accentuated in field studies. Here the AE/MS signals may be of relatively low frequency, and the "first break" (point of arrival) is spread out over several milliseconds. In other situations the background noise levels may be so high that the first arrival is obscured on one or more of the transducer channels. Experience to date has indicated that in such cases human judgement is invaluable in separating signals from noise and allowing realistic arrival time values to be determined.

Unfortunately, the increasing tendency to require on-line processing of large volumes of AE/MS data negates the use of manual procedures and research has been undertaken to develop more powerful, computer-based, procedures to generate arrival time data and carry out subsequent source location analysis. A number of techniques for improving the quality of arrival time data include those listed below.

a. *Improved threshold detection*: In many AE/MS location systems, for example the system shown earlier in Figure 5.16, a threshold circuit is used for each transducer and AE/MS sources are located by detecting signals which exceed the present threshold level. In such systems the detecting and locating probability is influenced to a great extent by the threshold level. A high threshold value (low sensitivity) results in poor delectability. However, too small a threshold (high sensitivity) reduces the location performance, since the number of spurious locations are increased by the processing of random combinations of low-level signals. A paper by Hamada (1976) considers this problem and develops a simple mathematical model for AE/MS amplitude and attenuation, and describes quantitatively detection probabilities and location performance.

b. *Cross correlation approach*: Cross correlation techniques can be used to measure the arrival-time-difference for both discrete and continuous AE/MS signals. Baron and Ying (1987) provide a brief review of this technique and emphasize its usefulness, particularly when dealing with continuous AE/MS signals, such as those related to fluid flow and leakage.

c. *ADASLS program*: ADASLS (Automatic Data Analysis and Source Location System), is a sophisticated software package developed at CANMET (Ge and Mottahed, 1993, 1994). The most powerful feature of this code is its ability to identify the type of arrival picks (i.e., P-wave, S-wave and outliers) in the absence of actual waveform data. This feature makes it possible to analyze AE/

MS field data on a scientific basis that is not possible using any conventional methods. The effectiveness of this feature was demonstrated at a number of mines, where in all cases, the source location accuracy of the data being evaluated was drastically improved. An important feature of ADASLS, which is critical for a reliable daily monitoring program, is the ability to assess the reliability of event data and event locations. The AE/MS data, especially those which are determined in an automated process, are very complicated. Furthermore, many of these are not real events, in that the minimum number of arrivals necessary for source location calculation are not available.

d. *Earthquake related techniques*: Lee and Stewart (1981) discuss a wide variety of visual, semiautomated and automated methods for earthquake event detection. A number of these methods, in particular, those based on the short-term/long-term signal averages have been successfully applied to the processing of AE/MS field data.

   Niitsuma and his associates, at Tohoku University in Japan, have developed a number of unique methods for optimizing source location accuracy for data obtained using single station seismometry. These methods will be considered briefly later in Section 5.10.2 of this chapter.

5.9.3   Transducer coordinates

Since the transducer coordinates enter directly into the source location calculations they should be as accurate as possible. The required degree of accuracy is, however, relative to the size of the structure involved and the associated transducer array. In laboratory studies transducers may be installed to within perhaps ± 1 mm. In the field, however, this degree of accuracy is not normally required and installation to within ±0.3 m to ±2.0 m may be adequate when dealing with large structures.

5.9.4   Transducer array geometry and size

The array geometry and the location of the event relative to the array play a major role in establishing the source location error. Each array has its own error boundaries and, consequently the suitability of a specific array should be investigated before it is utilized. Generally, as the source moves away from the center of the array, the source location error increases. Furthermore when a planar surface transducer array is utilized the vertical location error increases with the event depth (i.e., distance below the array plane).

   In general, the studies have indicated that the the size of the array influences the location error as follows:
1. Source locations may be very accurate for a small array, provided that they originate within the boundaries of the array.

2. Source locations are less accurate for a large array, however, in this case a greater area may be monitored.

Considerable research over the last 10 years has indicated the critical importance of transducer array geometry on the accuracy of AE/MS source location (Ge, 1988, Ge and Hardy, 1988; Ge and Mottahed, 1993, 1994). These studies are based on the fact that the governing equations associated with the arrival-time-difference approach represent hyperboloids, and that a study of the hyperbolic field allows the role of array geometry to be interpreted without the introduction of arbitrary assumptions. Associated studies have indicated that the mechanism of array geometry control of source location accuracy may be summarized as follows:

1. The array geometry itself does not induce any errors. It merely "amplifies" errors already present.
2. The essence of this amplification in terms of geometry is that the source is incorrectly located on an adjacent hyperboloid rather than on the one associated with the true source. Thus, the density of hyperboloids in the region of the true source is a measurement of potential source location accuracy.
3. The effect of array geometry on source location accuracy is a result of the non-uniformity of the hyperbolic field, which makes the location error heavily dependent on the position of the true source relative to the array.

Ge also developed a simulation procedure by which source location accuracy could be predicted for a specific field site and selected array geometry (Ge and Hardy, 1988). This makes it possible to compute the source location accuracy for a specific array and to select the optimum array geometry for each AE/MS field site prior to actual transducer installation. In this simulation process the mean and other statistical properties of the field velocity are utilized to compute the variation of the computed source location, in terms of the total confidence interval, as a function of the position of the true source location. This is computed on the basis of the type of source location analysis utilized (e.g. least squares form of three-dimensional arrival-time-difference method) and the statistical properties of the velocity model associated with the specific field site under study.

Based on an analysis of Ge (1988), Figure 5.17 illustrates the variation in total coincidence interval (DR) associated with the transducer array used at the New Haven, Michigan gas storage reservoir site. The data presented is based on the assumption that the major AE/MS activity occurs at the 274 m (900 ft) level. The figure illustrates the central study area and for the majority of this area $10 \le SD \le 20$. The figure also shows how the DR contour plots are crowded together near transducers PSU-1, PSU-4 and PSU-6, indicating the potential for high location errors in these areas.

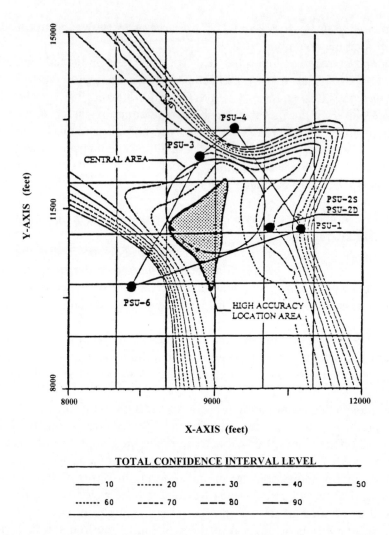

Figure 5.17. The total confidence interval associated with the original transducer array used at the New Haven Gas Storage Site (after Ge, 1988). (The monitoring area is assumed to be located at the 900 foot level. Solid circles denote transducer positions. Lowest level CD contours indicate regions where source location errors will be smallest. Conversion factor: 1 ft = 0.3048 m.)

5.9.5 Source location relative to array

Experience has shown that AE/MS source location accuracy decreases with distance from the array if the source is located outside the array. This effect is also clearly illustrated in model and field data provided by Leighton and Duvall (1972). See for example Figure 5.15b presented earlier.

5.10 OTHER SOURCE LOCATION TECHNIQUES

The majority of this chapter has focused on the use of multiple station seismometry and in particular the determination of source locations using arrival-time-difference procedures. This approach is most commonly utilized in geotechnical applications, however in certain special applications other techniques have been found to be more suitable. This section will examine a number of these, in particular zonal source location, single station seismometry, and hybrid techniques.

5.10.1 Zonal source location

a. *General*: This technique utilizes multiple station seismometry but analyzes the resulting AE/MS data in terms of the location of the "first-hit" transducer, the first transducer to detect the AE/MS signal; and the "hit-sequence," the time sequence of transducers (i.e. 1st, 2nd, etc.) which detect the AE/MS signal.

In the near-surface geotechnical application of AE/MS techniques, two factors in particular degrade the required accuracy of the input parameters necessary for utilization of conventional (arrival-time-difference) source-location techniques. There are the inability to select accurate wave arrival times and poorly defined velocity models. In such situations, conventional, least-squares based source-location procedures, which converge to precise location estimates, far exceed the levels of precision of the associated input parameters.

Similarly, in non-geologic materials, problems associated with high signal attenuation, variable velocity structures, and low signal-to-noise (S/N) ratios have also been encountered (Barsky and Hsu, 1985; Nakasa, 1984). As a consequence, so-called zonal-location methods have been utilized in the on-line monitoring of such structures as fiber-reinforced-plastic (FRP) and metal pressure vessels (Arrington, 1984; Fowler, 1984). Such location methods may apply equally well to near-surface geotechnical situations. Barron and Ying (1987) discuss procedures for utilizing both zonal and hit-sequence techniques.

Although several variations exist, zonal-location algorithms simply search for the transducer with the first signal arrival (hit) and immediately establish a primary zone of activity. Successive wave arrivals at additional transducer sites and, in some algorithms the arrival-time differences, are used to refine the zone of activity. The use of the arrival-time sequence and arrival-time differences to obtain a refined source location is referred to by some authors as hit-sequence location (HSL) (Fowler, 1984). Essentially the HSL method defines a smaller subzone within the original primary zone from which the event originated.

Figure 5.18 shows a simplified field situation illustrating the zonal and hit-sequence location concept. Here the circles denote an array of transducers (sensors), and the assumed transducer hit-sequence is 1, 2 and 3. Based on the

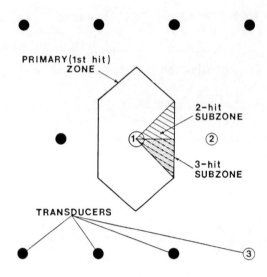

Figure 5.18. Simplified field situation illustrating the zonal and hit-sequence location concept.

simple zonal location technique the AE/MS source is assumed to be located in the primary (1st hit) zone surrounding the 1st hit transducer. Based on the HSL technique (2nd and 3rd hit transducers) the source is assumed to be located in a specific area of the primary zone, namely the 3-hit subzone.

Zonal location techniques offer four advantages over conventional, arrival-time difference source-location routines (Fowler, 1984), namely:

1. Severe attenuation, which often results in only a few transducers in a multi-transducer array detecting the AE/MS event (i.e. insufficient transducer hits) or necessitates impractical transducer spacing for conventional analysis, is accounted for in zonal methodology.

2. Approximate velocity models, which can deviate significantly from reality, due to local velocity variations, and can lead to erroneous solutions when utilizing conventional source location techniques, are not strictly necessary in zonal methodology.

3. Extensive computation time required for source location using conventional methods is reduced considerably using zonal methods, since these incorporate simpler source location algorithms. As a result, zonal location can often be performed in real time.

4. Zonal location algorithms make use of the majority of incoming data. In contrast, "valid events," based on an adequate number of transducer hits (e.g. five for three-dimensional situations) are required when utilizing conventional arrival-time-difference techniques. In many situations this requirement eliminates a significant portion of the available data.

Field experience in FRP and metal structural analysis has shown the zonal philosophy to be reliable and relatively independent of velocity errors and transducer placement (Fowler, 1984). Required input data for utilization of zonal-location techniques are the general attenuation and velocity characteristics of the structure. This data permits proper transducer placement and delimits and various zones (and subzones) within the monitoring area (Arrington, 1984).

b. *Typical field application*: In order to demonstrate the suitability of the zonal technique data from a recent Penn State AE/MS study, associated with the evaluation of methods for monitoring sinkhole development at a Pennsylvania airport site (Hardy et al., 1987), was analyzed. This allowed a comparison to be made of the source-location accuracy associated with the conventional arrival-time-difference technique, and the zonal and hit-sequence techniques. Commercially available geophones were utilized as AE/MS transducers. Figure 5.19 shows the seven-transducer array and calibration-source positions at one of the airport test sites. A steel-bar, having an attached plate in contact with

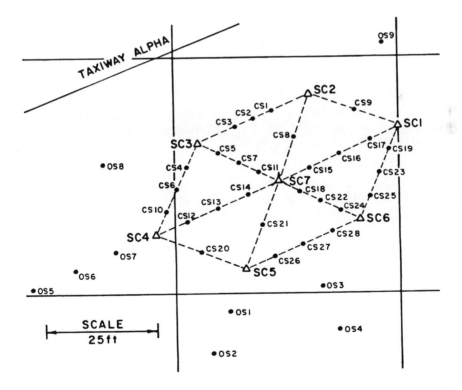

Figure 5.19. AE/MS - Transducer array and calibration-source point positions at airport test site no. 2 (after Hardy et al., 1987). (SC denotes transducer locations, CS and OS denote calibration-source points inside and outside the array, respectively. Length conversion factor: 1 ft = 0.305 m.)

the ground surface and impacted with a heavy hammer, was used as a seismic source. The wave generated was assumed to be an S-wave or pseudo-Raleigh wave. To obtain data for development of the required velocity model the seismic source was sequentially activated at each of the seven AE/MS-array transducer locations (SC1, SC2, ..., SC7) and arrival-time data monitored at each of the six remaining transducer locations. The global average velocity computed from the data was found to be 765 ft/sec (233 m/sec) with a standard deviation of 165 ft/sec (50 m/sec).

In order to evaluate source location accuracy a seismic source was activated at a number of accurately known positions throughout the AE/MS array. A "soft-source", consisting of a cloth sack of loose soil weighing approximately 5 kg (11 lb) was used in order to mimic a sinkhole-induced soil collapse. As indicated in Figure 5.19, twenty-eight calibration-source positions within the array (CS points) and ten calibration-source positions located outside of the array (OS points) were utilized. Arrival-time-difference data and the ordered sequence of transducers to receive a given signal (hit-sequence) were compiled for each calibration-source point (Hardy et al., 1986).

Source location at the airport site was considered to be a two-dimensional problem. A modified form of the USBM arrival-time-difference program was implemented for this analysis and an isotropic velocity model using the global value of 767 ft/sec (233 m/sec) computed earlier. Zonal and hit-sequence location techniques, discussed earlier were employed.

Figure 5.20 illustrates some typical results. In each case, the solid dot is the true position of the activity and the open circle is the conventionally located source position. In Case A, zonal location selects the area immediately surrounding transducer position SC1 (bold outline). Two-hit based hit sequence location (HSL) yields the crosshatched area representing the hit sequence SC1 followed by SC6. The three-hit HSL yield the shaded zone representing the hit-sequence of SC1, SC6, and SC7. In this case, zonal and HSL results coincide exactly with the known source position. In Case B, there is a good agreement between the true source position and that determined by conventional location procedures. Zonal location is also satisfactory but results obtained by two- and three-hit HSL are not as favorable as in case A. In Case C, conventional location yields very poor results but zonal and two-hit HSL results are excellent. Finally, in Case D, both conventional and three-hit sequence zonal locations agree well with the true source location.

Based on the analysis of data collected at the Capital City airport site, it appears that, although potentially less precise than conventional arrival-time-difference location, zonal location along with hit-sequence location can provide a meaningful indication of the active area in both space and time. Location of this area, rather than the spatial coordinates of individual instabilities, is all that is necessary in many geotechnical applications. Furthermore, it is important to note that in those cases where high attenuation and only a poorly defined

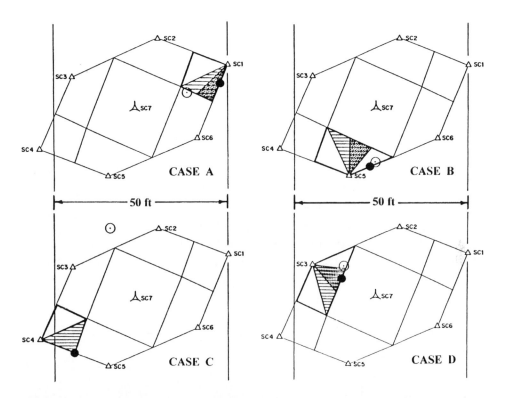

Figure 5.20. Comparison of computed source locations using conventional, zonal and hit-sequence techniques (after Hardy et al., 1987). (Triangles indicate transducer positions, solid and open dots indicate true source and conventionally computed source locations. Bold outlined regions indicate zonal location. Hatched and hatched-stippled regions indicate two- and three- hit hit-sequence locations. Length conversion factor: 1 ft = 0.305m.)

velocity model is available, the hit-sequence technique may, in fact, provide the only meaningful source location information.

5.10.2 Single station seismometry

a. *General*: Chapter 3, Section 3.5 briefly describes the development of single station seismometry systems for use in geothermal energy research. Such systems, often referred to as "sondes", employ a triaxial transducer system for detection and location of AE/MS sources. Early development and application of such systems at Los Alamos National Laboratory (LANL) are described in a paper by Albright and Pearson (1982). Figure 5.21a illustrates a simplified schematic of the LANL triaxial sonde and Figure 5.21b illustrates the location of the triaxial transducer and the related source location geometry. The exact coordinates of the associated triaxial transducer are determined using a series of explosions at known locations (Albright and Pearson, 1978) or by the use of a built-in electronic compass (Niitsuma et al., 1989).

When the transducer detects an AE/MS event, separate signals are generated by the vertical element (V) and the two horizontal elements (H1 and H2). These three signals are recorded, processed and combined to provide the three-dimensional location of the source. Figure 5.22 illustrates a set of typical signals. The initiation of both the P-wave and S-wave components are shown for the vertical signal.

b. *Source coordinates*: As shown in Figure 5.21b, in order to locate a specific source requires the spherical coordinates, R, θ and ϕ. These coordinates are often obtained using the following procedures:

Radius R: This factor is determined using the difference in arrival time of the P- and S-waves and the P- and S-wave velocities. The velocity data is normally obtained by monitoring data generated by seismic sources located at known locations. Equation 2.42, presented earlier, can then be applied, namely:

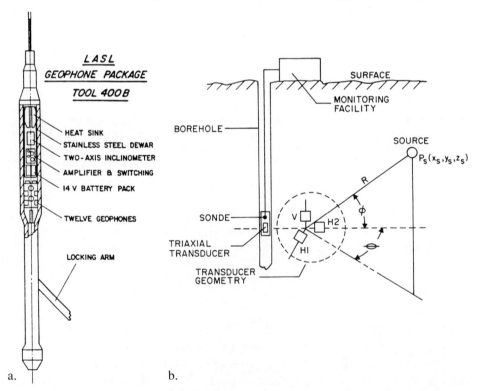

Figure 5.21. LANL triaxial sonde and related source location geometry. (V, H1 & H2 denote, respectively, the vertical and the two horizontal transducers. Actually the LANL system utilized a total of 12 transducers (geophones), four in each direction.) a. Simplified schematic of LANL sonde (after Pearson & Albright, 1983). b. Geometry associated with single station seismometry.

$$R = \left[\frac{C_1 C_2}{C_1 - C_2} \right] \Delta t$$

where C_1 and C_2 are the P- and S-wave velocities, and Δt is the difference in the P- and S-wave arrival times.

Angle θ: The azimuthal angle is obtained by a procedure (Lissajous figure) similar to that used to evaluate the phase shift between two electrical signals (e.g. see Doebelin, 1975). The procedure involves cross-plotting the data obtained from the two horizontal transducers (H1 and H2) as shown in Figure 5.23. An elliptical figure is obtained and the orientation of the major axis of the ellipse relative to the H2 axis (y-axis) provides the value of θ. This analysis procedure is denoted as the Hodogram technique. It is based on the fact that for a compressional wave, the first particle motion is in the direction of propagation so that the determination of azimuthal approach of the signal, gives the direction of the vector (R) from the detector (transducer) to the event source.

Angle ϕ: The angle of inclination ϕ is determined by applying the hodogram technique to the V/H1 and the V/H2 data sets. If the associated data is consistent, both sets of data will give the same ϕ value, otherwise the mean of the two values should be used.

c. *Other analytical procedures*: If sufficient computer capability is available, data from triaxial transducers may be analyzed directly to provide source location data. Such analysis is based on the procedures outlined earlier in Chapter 2 (Section 2.9.5).

A number of special procedures for analysis of data obtained using single station seismometry have been developed by Professor Niitsuma, Tohoku University, Japan and a number of his associates. These include source location procedures using spectral matrix analysis (Moriya et al., 1990), accurate calibration of downhole monitoring systems (Sato et al., 1986), doublet/ multiplet analysis (Moriya and Niitsuma, 1998), and the use of "crack wave" particle motion to determine fracture orientation (Nagano et al., 1998).

5.10.3 Hybrid techniques

a. *Western deep levels studies*: Rock bursts are a major problem in most deep hardrock mines. During the late 1970s and early 1980s a sophisticated AE/MS monitoring system was developed at the Western Deep Levels gold mine, near Carletonville, South Africa. Initial studies (Van Zyl Brink and D. O'Connor, 1984) involved the use of accelerometer-type triaxial transducers, installed underground at a depth of greater than 3000 m. These studies utilized single station seismometry techniques to monitor a specific mining face. The azimuth and elevation of the source relative to the transducer was determined by a least squares comparison of the individual components. Distance to the

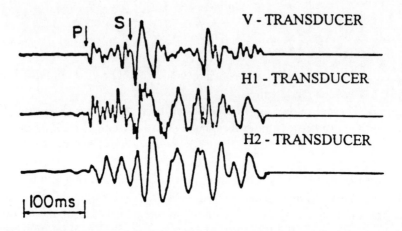

Figure 5.22. Set of typical AE/MS signals detected by a triaxial sonde (after Albright and Pearson, 1979).

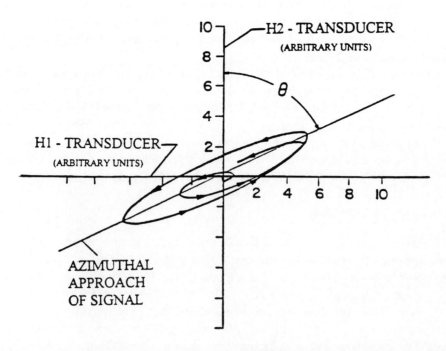

Figure 5.23. Hodogram of particle motions detected by two horizontal transducers, H1/H2. (Figure illustrates the method of determining azimuthal angle θ.)

source was estimated in the conventional manner using the P-S wave arrival-time-difference. However, coordinate rotation was first employed to resolve the longitudinal P-wave from the transverse S-wave.

Following the successful initial studies a larger extremely sophisticated facility was developed at the Western Deep Level mine (Brink and Mountfort, 1984). This included both a microseismic and a macroseismic system. The microseismic system included five triaxial accelerometer-type transducers each having a dedicated minicomputer for event detection, location and storage. The macroseismic system included twenty-four triaxial geophones and a single dedicated minicomputer. Data from all transducers was transmitted by cable to a surface monitoring facility where the dedicated minicomputers and an associated supervisory computer were located. The facility is clearly hybrid in form, consisting of an array of transducer sites with each transducer being triaxial in form. This allows the location of AE/MS events using both single and multiple station seismometry. This facility is capable of detecting AE/MS events over a large monitoring area, and over a wide range of magnitudes.

b. *Camborne studies*: For some years studies have been underway to study hydraulic stimulation of a potential "hot dry rock" (HDR) geothermal reservoir. An AE/MS project, associated with these studies, was undertaken in Cornwall, England, in an effort to map the locations of AE/MS activity generated during hydraulic stimulation experiments. These experiments were carried out in an effort to interlink two boreholes, an injection and a production well, previously drilled some 2000 m into the Carmenellis granite pluton. Technical and economic restraints prevented the use of conventional single station or multiple station seismometry. Instead, a hybrid system was utilized, consisting of a string of three hydrophones located in the production well, four remote stations each consisting of a vertical transducer grouted at the bottom of a 200 m deep-borehole, and a triaxial transducer mounted on a granite outcrop near the test site. Data from all transducers was initially processed and then recorded on a multi-channel FM recorder. Off-line, the recorded data was digitized and recorded on disk using a suitable minicomputer. The stored data was later analyzed using the computed arrival-time-difference data, and both the conventional least squares method and the Seven Point Approximation Method (SPAN), Dechman and Sun (1977).

REFERENCES

Albright, J.N. and Pearson, C.F. 1979. *Microseismic Monitoring at Byran Mound Strategic Petroleum Reserve*, Internal Report, Los Alamos Scientific Laboratory, Los Alamos Scientific Laboratory, Los Alamos, March 1979.

Albright, J.N. and Pearson, C.F. 1982. Acoustic Emissions as a Tool for Hydraulic Fracture Location: Experience at the Fenton Hill Hot Dry Rock Site, *Society Petroleum Engineering Journal*, August 1982, pp. 523-530.

Arrington, M. 1984. In-situ Acoustic Emission Monitoring of a Selected Mode in an Offshore Platform, *Proceedings 7th International Acoustic Emission Symposium*, Zao, Japan, Japanese Society for Nondestructive Inspection, Tokyo, pp. 381-388.

Asty, M. 1978. Acoustic Emission Source Location on a Spherical or Plane Surface. *Non-Destructive Testing International*, October 1978, pp. 223-226.

Baria, R.K. and Batchelor, A.S. 1989. Induced Seismicity During the Hydraulic Stimulation of a Potential Hot Dry Rock Geothermal Reservoir, *Proceedings Fourth Conference on Acoustic Emission/Microseismic Activity in Geologic Structures and Materials*, The Pennsylvania State University, October 1985, Trans Tech Publications, Clausthal-Zellerfeld, Germany, pp. 327-352.

Baron, J.A. and Ying, S.P. 1987. Acoustic Emission Source Location, *Nondestructive Testing Handbook*, Vol. 5: Acoustic Emission Testing, Section 6, P. M. McIntire (ed.), American Society for Nondestructive Testing, Columbus, Ohio, pp. 135-153.

Barsky, M. and Hsu, N.N. 1985. A Simple and Effective Acoustic Emission Source Location System, *Materials Evaluation, Vol. 43*, pp. 108-110.

Blake, W. 1971. *Rock Burst Research at the Galena Mine, Wallace, Idaho*, TPR 39, U.S. Bureau of Mines.

Blake, W. and Duvall, W.I. 1969. Some Fundamental Properties of Rock Noises, *Trans. AIME, Vol. 244*, No. 3, pp. 288-290.

Blake, W. and Leighton, F. 1970. Recent Developments and Applications of the Microseismic Method in Deep Mines, *Rock Mechanics Theory and Practice*, Proceedings, 10th Symposium on Rock Mechanics, editor: W. H. Somerton, AIME, New York, pp. 429-443.

Blake, W., Leighton F. and Duvall, W.I. 1974. Microseismic Techniques for Monitoring the Behavior of Rock Structures, *USBM Bulletin 665*, 65 pp.

Brink, A.V.Z. and Mountfort, P.I. 1984. Feasibility Studies on the Prediction of Rockbursts at Western Deep Levels, *Proceedings 1st International Congress on Rockbursts and Seismicity in Mines*, Johannesburg, South Africa, South African Institute Mining and Metallurgy, Symposium Series No. 6, pp. 317-325.

Brink, A.V.Z. and O'Connor, D. 1984. Rock Burst Prediction Research–Development of a Practical Early-Warning System, *Proceedings Third Conference on Acoustic Emission/Microseismic Activity in Geologic Structures and Materials*, The Pennsylvania State University, October 1981, Trans Tech Publications, Clausthal-Zellerfeld, Germany, pp. 269-282.

Byerlee, J.D. and Lockner, D. 1977. Acoustic Emission During Fluid Injection, *Proceedings First Conference on Acoustic Emission/Microseismic Activity in Geologic Structures and Materials*, Pennsylvania State University, June 1975, Trans. Tech. Publications, Clausthal, Germany.

Cook, N.G.W. 1963. The Seismic Location of Rockbursts, Roch Mechanics, C. Fairhurst (ed.), *Proceedings Fifth Symposium on Rock Mechanics*, Pergamon Press, New York, pp. 493-516.

Dechman, G.H. and Sun, M.-C. 1977. Iterative Approximation Techniques for Microseismic Source Location, *USBM R.I. 8254*, 23 pp.

Doebelin, E.O. 1975. *Measurement Systems*, McGraw-Hill Book Company, New York, pp. 591-593.

Fowler, T.J. 1984. Acoustic Emission Testing of Chemical Process Industry Vessels, *Proceedings, 7th International Acoustic Emission Symposium*, Zao, Japan, Japanese Society for Nondestructive Inspection, Tokyo, pp. 421-449.

Ge, M. 1988. *Optimization of Transducer Array Geometry for Acoustic Emission/Microseismic Source Location*, Ph.D. Thesis, The Pennsylvania State University, Department of Mineral Engineering, August 1988.

Ge, M. and Hardy, H.R. Jr. 1988. The Mechanism of Array Geometry in the Control of AE/MS Source Location Accuracy, *Proceedings 29th U.S. Symposium on Rock Mechanics*, Minneapolis, MN, A. A. Balkema, Rotterdam, pp. 597-605.

Ge, M. and Mottahed, P. 1993. An Automatic Data Analysis and Source Location System, *Proceedings 3rd International Symposium on Rockbursts and Seismicity in Mines*, Kingston, Ontario, A. A. Balkema, Rotterdam, pp. 343-348.

Ge, M. and Mottahed, P. 1994. An Automated AE/MS Source Location Technique Used by Canadian Mining Industry, *Proceedings 12th International Acoustic Emission Symposium*, Sappora, Japan, Japanese Society for Nondestructive Inspection, Tokyo, pp. 417-424.

Hamada, T. 1976. Effects of Threshold Level on Probabilities of Detection and Location of AE Sources, *Proceedings 3rd International Acoustic Emission Symposium*, Tokyo, Japan, September 1976, Japanese Society for Nondestructive Inspection, Tokyo, pp. 101-118.

Hardy, H.R., Jr. 1986. Source Location Velocity Models for AE/MS Field Studies in Geologic Materails, *Proceedings 8th International Acoustic Emission Symposium*, Tokyo, Japan, October 1986, Japanese Society for Nondestructive Inspection, pp. 365-388.

Hardy, H.R. and Ge, M. 1986. A Computer Software Package for Acoustic Emission Field Application, *Proceedings 27th Symposium on Rock Mechanics*, AIME, New York, pp. 134-140.

Hardy, H.R. Jr., Belesky, R.M. and Ge, M. 1987. Acoustic Emission/Microseismic Source Location in Geotechnical Applications, *Proceedings, 4th European Conference on Non-Destructive Testing*, London, September 1987, Pergamon Press, New York, pp. 3066-3075.

Hardy, H.R., Jr., Mowrey, G.L. and Kimble E.J. Jr. 1978. Microseismic Monitoring of a Longwall Coal Mine: *Volume I – Microseismic Field Studies, Final Report, USBM Grant No. G0144013*, Pennsylvania State University, pp. 318.

Hardy, H.R. Jr., Belesky, R.M., Mrugala, M., Kimble, E. and Hager, M. 1986. *A Study to Monitor Microseismic Activity to Detect Sinkholes*. Federal Aviation Administration, U.S. Dept. of Transportation, DOT/FAA/PM-86-34, National Technical Information Service.

Kijko, A. 1975. An Algorithm and a Program for Foci Location in the Region of the Upper Silesia, Seminar on Mining Geophysics, Mining Institute of Czechoslovakian Academy of Sciences, Prague, November 1973, *Acta Montana IV (32)*, pp. 61-75.

Lee, W.H. and Stewart, S.W. 1981. *Principles and Applications of Microearthquake Networks*, Academic Press, New York, pp. 50-70.

Leighton, F. and Blake, W. 1970. *Rock Noise Source Location Techniques*, U.S. Bureau of Mines, RI 7432.

Leighton, F. and Duvall, W.I. 1972. *A Least Squares Method for Improving the Source Location of Rock Noise*, U.S. Bureau of Mines, RI 7626.

Miller, W. and Harding, S.T. 1972. Error Analysis of a Five-Station P-Wave Location Technique, *Bulletin Seismological Society America, Vol. 62, No. 4*, pp. 1073-1077.

Moriya, H., Nagano, K. and Niitsuma, H. 1990. Precise Estimation of AE Source Direction by Spectral Matrix Analysis, *Proceedings 10th International Acoustic Emission Symposium, Progress in Acoustic Emission V*, Japanese Society for Nondestructive Inspection, Tokyo, pp. 244-251.

Moriya, H. and Niitsuma, H. 1998. Doublet Analysis for Characterizing Regional Structures in Three-Component Microseismic Measurement, *Proceedings Sixth Conference on Acoustic Emission/Microseismic Activity in Geologic Structures and Materials*, Penn State University, June 11-13, 1996, Trans Tech Publications, Clausthal-Zellerfeld, Germany, pp. 401-410.

Mowrey, G.L. 1977. Computer Processing and Analysis of Microseismic Data, Proceedings First Conference on Acoustic Emission/Microseismic Activity in Geologic Structures and Materials, Pennsylvania State University, June 1975, *Trans. Tech. Publications*, Clausthal, Germany.

Nagano, K., Sato, K. and Niitsuma, H. 1998. Fracture Orientation Estimated from 3C Crack Waves, *Proceedings Sixth Conference on Acoustic Emission/Microseismic Activity in Geologic Structures and Materials*, The Pennsylvania State University, June 1996, Trans Tech Publications, Clausthal-Zellerfeld, Germany, pp. 411-419.

Nakasa, H. 1984. Some Critical Remarks on Industrial Applications of Acoustic Emission, *Proceedings 7th International Acoustic Emission Symposium*, Zao, Japan, Japanese Society for Nondestructive Inspection, Tokyo, pp. 286-293.

Nittsuma, H., Nakatsuka, K., Chubachi, N., Yokoyama, H. and Takanohaski, M. 1989. Downhole AE Measurement of a Geothermal Reservoir and its Application to Reservoir Control, *Proceedings Fourth Conference on Acoustic Emission/ Microseismic Activity in Geologic Structures and Materials*, The Pennsylvania State University, October 1985, Trans Tech Publications, Clausthal, Germany, pp. 475-489.

Pearson, C. and Albright, J.N. 1983. Acoustic Emission During Hydraulic Fracturing Experiments, *Proceedings Third Conference on Acoustic Emission/Microseismic Activity in Geologic Structures and Materials*, The Pennsylvania State University, October 1981, Trans Tech Publications, Clausthal, Germany, pp. 559-575.

Rindorf, H.J. 1981. Acoustic Emission Source Location, Bruel & Kjaer, *Technical Review* No. 2-1981.

Rothman, R.L. 1975. Crack Distribution Under Uniaxial Load and Associated Changes in Seismic Velocities Prior to Failure, PhD. Thesis, Geophysics Section, Department of Geosciences, The Pennsylvania State University.

Salamon, M.D.G. and Wiebols, G.A. 1974. Digital Location of Seismic Events by an Underground Network of Seismometers Using the Arrival Times of Compressional Waves, *Rock Mechanics, Vol. 6*, pp. 141-166.

Sato, M., Nakatsuka, K., Niitsuma, H. and Yokoyama, H. 1986. Calibration of Downhole AE Measuring Systems by Detonation Test, *Proceedings of 8th International Acoustic Emission Symposium, Progress in Acoustic Emission III*, Japanese Society for Nondestructive Inspection, Tokyo, pp. 389-395.

Tobias, A. 1976. Acoustic Emission Source Location in Two-Dimensions by an Array of Three Sensors. Non-Destructive Testing, February 1976, pp. 9-12.

CHAPTER 6

Special topics

6.1 INTRODUCTION

Chapters 3 and 4 provide a relatively comprehensive discussion of AE/MS field and laboratory techniques, and the associated instrumentation. However, a number of special topics, not discussed in detail earlier, and considered useful to the reader are included here. These topics include material on the following: geophone calibration, monitoring facility optimization and calibration, velocity measurement, waveguides, amplitude distribution analysis, pattern recognition, source parameter analysis, moment tensor analysis, Kaiser effect studies, and related NDT techniques. Although the listed topics are not considered in depth, suitable references are included to allow the reader to obtain more detailed information.

6.2 GEOPHONE CALIBRATION

6.2.1 General

In the general area of geotechnical engineering, especially in field applications, relatively low-frequency transducers, namely geophones, are normally employed as transducers due to the high attenuation of the high-frequency components of the associated AE/MS signals. Past experience indicates that the calibration data supplied with new geophones often lacks sufficient detail. Furthermore, geophone retrieved from long-term underground installation need to be recalibrated to determine their current characteristics and performance capabilities if they are to be utilized in future studies. A detailed review of geophone calibration procedures are presented by Oh (1996).

To calibrate a vibration transducer, such as a geophone, it is necessary to accurately determine its sensitivity at various frequencies of interest. The "back-to-back" comparison method appears to be a convenient technique compared to the "absolute" calibration method. The former method involves coupling the test transducer to a standard transducer (usually an accelerometer) and driving the pair with a vibration exciter at various frequencies and various acceleration or velocity levels. It is assumed that since the two transducers are coupled tightly together, both will experience exactly the same motion, and the behavior of the test transducer can be compared to that of the standard transducer.

6.2.2   Calibration system

A block diagram of the calibration system, recently developed by Oh at Penn State (Oh, 1996, 1998), is presented in Figure 6.1. The system includes a B&K Type 2034 dual channel signal analyzer, a PC and GPIB interface card, a B&K Type 4808 vibration exciter, a B&K Type 4379 reference accelerometer mounted on the test fixture, a B&K WB 0814 programmable attenuator, and other associated instruments. The signal analyzer is a fully self-contained two-channel FT analysis system which can measure and display the data in terms of 34 different time domain, frequency domain, and statistical functions. One of these functions, the frequency response function (transfer function)

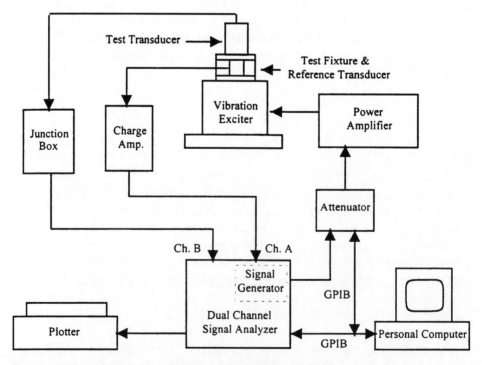

Figure 6.1. Block diagram of geophone calibration system (after Oh, 1998).

is used to measure the ratio of output voltages from the test and reference transducers, and the instantaneous spectrum is used for the driving the velocity level adjustment program. A built-in generator is used to supply a sine wave or random noise as an input signal to the vibration exciter. The signals generated by the analyzer are fed to the exciter through the signal attenuator and a power amplifier. The analyzer can be controlled by a personal computer through an IEC 625-1/IEEE 488 interface so that an operator-unattended test can be performed using an associated computer program.

The reference transducer used is an accelerometer. The calibration data for this transducer is traceable to the National Institute of Standards and Technology (NIST). Since the geophone output voltage is proportional to the velocity of excitation, the output signal from the reference accelerometer is integrated by an associated charge amplifier to obtain the excitation velocity. The calibrated output sensitivity is 1 volt/meter/sec and the operating frequency range is 2 Hz-10 Khz. The output from the reference accelerometer is also used for adjustment of the driving velocity.

The vibration exciter consists of permanent magnet and driving coil mounted at the bottom end of the moving table. The exciter is driven by an associated power amplifier. To maintain constant velocity, the amplitude of the input signal is set by an attenuator which is controlled by the computer program, using a signal from the reference transducer in a feed-back mode. Although vibration exciters are basically highly non-linear, using a developed computer program, excitation velocities in the range 0.0003 m/s (0.0118 in./s) to 0.003 m/s (0.118 in./s) varied less than ±6% over the full frequency range.

Test fixtures were designed to fasten the reference and test transducers to the vibration exciter. Since the geophone models studied varied in size and weight, two different test fixtures were designed, and fabricated from aluminum. A special fixture to excite the test geophone at various angles with respect to the vertical axis was also developed for use with the small test geophones. This allowed the geophone to tilt at any angle with the center of gravity of the test geophone remaining on the axis of excitation.

6.2.3 Calibration program

The Penn State Geophone Calibration Program (PSGEOCAL) was developed to automate the geophone calibration routines through the GPIB interface system. Details of the program, which is menu driven, are given elsewhere (Oh, 1996). Using the calibration program the system is able to carry out two different types of calibration, namely: the sinusoidal swept-frequency test ("Step-Sine Test") and the "Random Noise Test". Studies were carried out to compare the two calibration types. Generally, the test data for small 8 Hz and 14 Hz geophones, and for a large 1 Hz geophone, were in good agreement. Based on running time, the random noise test is the most cost effective, requiring approximately 0.25 hr compared with many hours for the step-sine test.

6.2.4   Test results

The completed calibration system was utilized to study the operating characteristics of a series of small and large geophones. Using either the step-sine or random noise calibration, three different sets of data are obtained from a single calibration test. These are the measured sensitivity of the test geophone, the phase angle difference between the reference transducer and test geophone, and the coherence function between the two transducers which indicates the "quality of the calibration". Typical test results are shown in Figure 6.2. Studies indicated that similar results were obtained by the step-sine and the random noise tests, and over a specific range the results were independent of driving velocity.

It is normally assumed that a batch of geophones fabricated to have a specific frequency response will have the same overall characteristics. However, small differences in materials and fabrication of certain components may result in a variation of frequency response between units. To investigate the degree of this variation, six 8 Hz and six 14 Hz, GSC-11D geophones were tested using the step-sine method and an excitation velocity of 0.118 in./s. The 8 Hz geophones, were tested over the range 5 to 400 Hz. Over this range the averaged sensitivity was found to be within -2% to 12% (-0.1 to +0.9 dB) of the theoretical value. Over the flat response region the deviation was less than 1% (0.1 dB). The 14 Hz geophones were tested over the range 8 to 800 Hz. Over this range the average sensitivity varied from -6% to +5% (-0.35 dB to +0.35 dB). Over the flat response region the deviation was less than 1% (0.1 dB). Clearly, for the geophones tested, there was little deviation between geophones in a specific batch. This indicates a high degree of quality control.

In AE/MS field studies, geophones are assumed to be installed in a vertical position. However, as shown in Figure 6.3a, in some cases the geophones may be installed at other angles, depending on the conditions at the installation site. Since the actual response of the geophone is dependent on the installation angle, studies to investigate this were carried out. In these studies a number of 8 Hz and 14 Hz, GSC-11D geophones were calibrated using a specially designed test fixture. After the test fixture was mounted on the exciter, the frequency response of the geophone in the vertical direction, was obtained. The geophone was then tilted to a specific angle (γ) and another calibration was carried out. The test angles were increased from 0° (vertical) up to 45° in increments of 5° or 10° for the 8 Hz geophones, and up to 80° in increments of 10° or 20° for the 14 Hz geophones.

Figure 6.3b shows typical results for a 14 Hz geophone with no shunt resistor. As seen in the figure, the measured sensitivity gradually deceased as

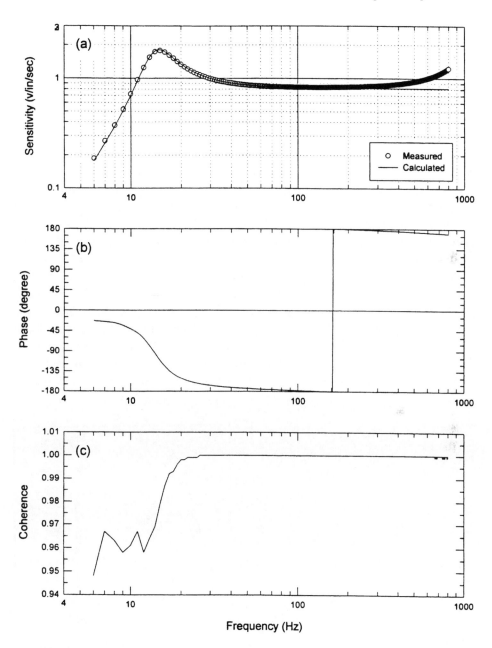

Figure 6.2. Typical results obtained from random noise calibration of geospace GSC-11D, 14 Hz test geophone: (a) Frequency response curve, (b) Phase angle, and (c) coherence (after Oh 1996). (Geophone No.14-1 with no stunt resistor. The apparent phase shift from -180° to 180° at a frequency of approximately 170 Hz is virtual and due only to the characteristics of the signal analyzer.)

the excitation angle increased up to 70° while generally maintaining the overall shape of the curves. For excitation angles greater than 60°, the overall shape of the measured sensitivity curves deviated greatly from the trend seen at smaller angles. In particular, note the instability observed at SF2. Behavior above 60° is considered unacceptable. Similar behavior was observed with different shunt resistors and with other test geophones. The well defined peaks (SF1) in the response curves in the range 150 Hz to 170 Hz were considered to be spurious geophone frequencies related to the characteristics of the suspension. In general, results indicated that the geophones studied could be used at off-vertical orientations, namely: up to 35° for 8 Hz and 60° for 14 Hz geophones. It should be reiterated, however, that when a geophone is not mounted vertically, the frequency response will change, possibly narrowing the useful operating frequency range and resulting in the development of a variety of spurious frequencies.

Normally geophones are installed at a field site for period of time varying from a few days to years. In the past little or no information has been available in regard to possible geophone deterioration due to extended field service. To investigate this further, three used and three unused 8 Hz and 14 Hz geophones were calibrated. The step-sine procedure was employed, and various driving velocities, ranging from 0.0003 to 0.003 m/s were employed. The used geophones had been installed underground at a field site for at least 2

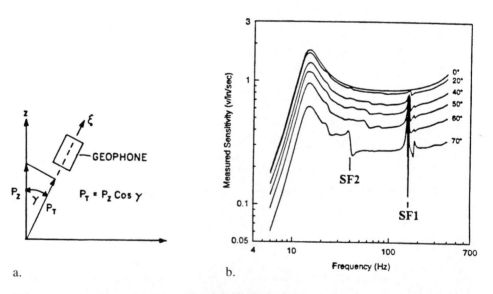

a.   b.

Figure 6.3. Details and results for transducer orientation tests on a GSC-11D, 14 Hz geophones (after Hardy and Oh, 1998). (Random noise method, no geophone shunt resistor. SF denotes spurious frequencies.) a. Relation of geophone axis (ξ) to shaker axis (Z) during tests. b. Frequency response curves obtained at various excitation angles (γ).

years. Although there were some minor differences in the characteristics of the used and unused geophones, it was difficult to determine if this was due to innate differences between similar types of geophones, as noted in the previous section, or due to actual field degradation. It is the author's opinion that little or no actual degradation occurred.

6.3   MONITORING FACILITY OPTIMIZATION

6.3.1   Introduction

The ability of laboratory and field AE/MS monitoring facilities to detect and process meaningful signals necessitates that the overall monitoring facility be optimized to provide a suitable signal-to-noise ratio (SNR). For example, Figure 6.4 illustrates a block diagram of a simple AE/MS field monitoring facility illustrating a variety of potential noise sources which need to be reduced or eliminated to obtain the desired SNR. These include various mechanical sources, which couple directly into the transducer, and a variety of internal and external electrical sources, which couple into the transducer, field cables and the monitoring system itself.

Mechanical sources: A wide range of mechanical sources exist. These include local and distant seismic activity; microseisms due to large scale sources such as ocean and earth tides; cultural activity due to the operation of

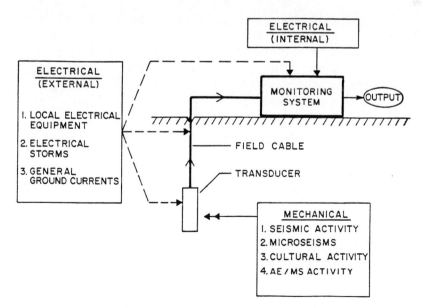

Figure 6.4. Simplified AE/MS field monitoring facility illustrating potential noise sources.

industrial facilities, highway, railway and air transportation, and nearby human and animal activity; and AE/MS activity due to the facilities under study and directly associated mechanical activities. Reduction of undesirable mechanical noise is most easily accomplished by monitoring the diurnal variations in background noise and limiting research monitoring to periods of low mechanical background. Another approach is to study the frequency character of the undesired mechanical noise and introducing electrical filters that can be set to reject such noise.

Electrical sources: Noise due directly to electrical sources include signals associated with local electrical equipment, electrical storms, ground currents, poor grounding and shielding techniques, and power line fluctuations and transients. Such noise may enter the monitoring facility at a number of sites, including the transducer, the field cables and the monitoring system itself. Electrical noise due to some sources, e.g. local electrical equipment and electrical storms, may be minimized by monitoring only at times when such sources are at a minimum. Generally electrical noise, as in the case of mechanical noise, may be significantly reduced by suitable electronic filtering and/or some other form of suitable signal discrimination.

A number of techniques for monitoring facility optimization, including those associated with transducer characteristics and placement, grounding and shielding techniques, monitoring system power, electronic filtering, and signal discrimination, will be discussed in the following sections.

6.3.2   Transducer characteristics and placement

Obviously it is desirable to utilize transducers that are most sensitive in the frequency range of the major AE/MS activity generated by the process under study. As the frequency of the major activity increases from low to high, the most suitable transducer type varies from displacement transducer to velocity transducers (geophones), to accelerometers and finally to AE-transducers. For example, if the major activity was in the range 20Hz – 500 Hz, a velocity transducer would be most suitable. In contrast, activity in the range 75kHz - 250 kHz would be best monitored by a suitable AE-transducer.

Since AE/MS signal amplitude is highly attenuated in geologic materials, the best SNR will be obtained when the transducer is located as close as possible to the source. However, it is important to remember that AE/MS source location accuracy depends on transducer array geometry and this may limit extensive minimization of the transducer-to-source distance.

6.3.3   Grounding and shielding techniques

Proper grounding and shielding of electronic components, transducers and connecting cables is extremely important in AE/MS monitoring facilities, particularly in field related studies. Optimum techniques will depend on the monitoring facility configuration. A number of texts are available dealing specifically with grounding and shielding techniques (e.g. Morrison, 1967). At Penn State considerable attention has been given to the problem relative to the development of an optimized monitoring system for use in AE/MS studies of gas storage reservoir stability (Hardy et al., 1981). Figure 6.5 illustrates the grounding and shielding system employed.

6.3.4   Monitoring system power

Undesirable electrical noise can also enter a highly sensitive AE/MS monitoring system through the facilities used to provide AC power to the system. This problem is accentuated in AE/MS field studies where potentially unfavorable power line conditions (voltage fluctuations, transients, etc.) may exist. At Penn State this problem was resolved by using a commercial AC power-line regulator unit (Topaz Model 3203-2) between the local power line and the AE/MS monitoring system (Hardy et al., 1981). The improvement using the regulator is illustrated by fact that the background noise level for the various transducer channels decreased from >300 mv rms to 10-90 μv rms with the introduction of the regulator unit.

6.3.5   Electronic filtering

Although "filtering" techniques can, in general, be based on a wide variety and combinations of parameters (e.g. sophisticated pattern recognition procedures)

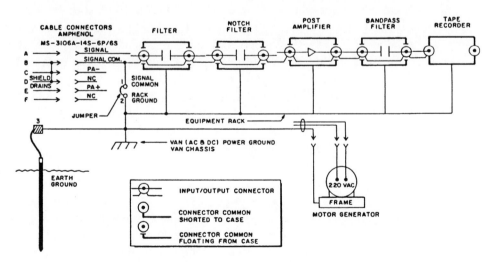

Figure 6.5. Simplified block diagram of grounding and shielding technique used in Penn State Mark II AE/MS field monitoring system (after Kimble, 1989).

in the present context, filtering will refer to the reduction or elimination of electrical signals by limiting the frequency response of the monitoring or subsequent signal processing facilities. In extremely noisy environments, such as a typical geotechnical site, extensive on-line filtering must be accomplished prior to data recording, otherwise the generally small AE/MS signals become hopelessly buried in background noise, and system saturation may occur. In the last 25 years, the author has had extensive experience in monitoring of AE/MS activity from surface at a variety of field sites. These include gas storage reservoirs, coal mines, surface and underground limestone mines, sinkholes and highway slopes.

Penn State studies in recent years have employed the Mark II AE/MS monitoring system described earlier in Section 3.6.4 (Kimble, 1989). In order to obtain meaningful data from typical geotechnical sites has required the development of extensive on-line filtering techniques. The first stage in this study involved the detailed investigation of the characteristics of the basic monitoring system itself. This was accomplished by the removal of the AE/MS transducers on two-channels, and placement by 910 Ω shunt resistors across the preamplifier inputs. During the study the output of the system was recorded on a built-in multichannel magnetic type recorder, and later played out in hard-copy, using an associated UV-recorder.

Using a pass-band of 10 Hz to 250 Hz and a system gain of 126 dB (approximately 2×10^6 voltage gain), the average noise signal for the two channels was found to be 1.1 volts peak-to-peak. Therefore the basic system referred to the noise input was 5.5×10^{-7} volts peak-to-peak, or 1.8×10^{-7} volts RMS. For a bandpass of 10 – 250 Hz (240 Hz width), the system noise level was computed to be ne = 8.9×10^{-9} volts RMS/Hz. Since the response of the geophones commonly used in AE/MS field studies were approximately 23.6 volts/m/sec, the equivalent ground motion noise was computed to be ng = 0.377×10^{-9} m/sec/Hz, which is considered excellent. It must be remembered, however, that the computed value of ng was obtained under optimum conditions. Under actual field conditions, AE/MS transducers are connected to the system inputs, and a noise figure somewhat higher would be expected due to ambient electrical and mechanical noise. Noise at any new field site must be studied in detail early in the project and optimum filter settings selected for routine monitoring.

As shown in Figure 6.6 the Mark II monitoring system was designed to allow the incorporation of two active band-pass filters and one notch filter in each monitoring channel. The low-pass and high-pass cut-off frequencies of each band-pass filter were switch selectable and their characteristics are shown in Figure 6.7a. To improve the system's ability to reject signals originating from 60 Hz power line signals and their harmonics, the notch filters were designed to reject frequencies of 60 Hz, 120 Hz, 180 Hz, and 540 Hz. Any combination of these frequecies were switch selectable, and the choice of these frequencies was based on past field experience. Figure 6.7b shows the frequency response of the system with the notch filters included. Further details on the design and construction of these filters are given in Kimble (1983). For those interested in the design of special filters, a number of books are available (e.g., Helburn and Johnson, 1983).

To further illustrate the importance of electronic filtering, the results of facility optimization at the New Haven gas storage site (Hardy and Mowrey, 1981) are included here. Based on a signal-to-noise ratio (SNR) of 1.5 and a transducer sensitivity of 295.2 volts/m/sec. the useable sensitivity of the preliminary monitoring system, which incorporated a single bandpass filter (1-40 Hz) and had a total system gain of 50-60 dB, ranged from $2.79\text{-}7.62 \times 10^{-6}$ m/sec. After incorporating an additional bandpass filter (1-40 Hz) and a 60 Hz notch filter, it was possible to raise total system gains to 89-90 dB, and resulting useable system sensitivities were increased to $2.79\text{-}8.38 \times 10^{-8}$ m/sec., depending on the particular transducer involved.

6.3.6 Spatial Discrimination

Early papers by Keledy (1975) and Nakamura (1974) discuss in some detail the problems of detecting characteristically weak AE/MS signals in a noisy electrical and mechanical environment. Keledy (1975) considers that signal discrimination techniques can be categorized as those based on (1) discrimination by analog signal properties, and (2) spatial discrimination. The former is based on properties such as amplitude, frequency, rise time, and known time of occurrence (e.g. during certain stages of a controlled loading cycle). In this case discrimination of "undesired signals" may be accomplished by suitable signal processing such as filtering. In contrast, spatial discrimination involves the use of a number of transducers and criteria associated with the sequence of signal arrivals at the various transducers.

Hardware-based spatial discrimination in one-and two-dimension situations is clearly illustrated by Pollock (1979) in a Dunegan/Endevco

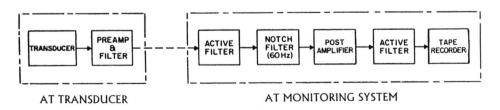

AT TRANSDUCER   AT MONITORING SYSTEM

Figure 6.6. Block diagram showing arrangement of electronic components in Penn State Mark II AE/MS monitoring system (after Hardy, et al., 1981). (A 60 Hz notch filter is shown but any combination of 60, 120, 180 and 540 Hz notch filters could be used.)

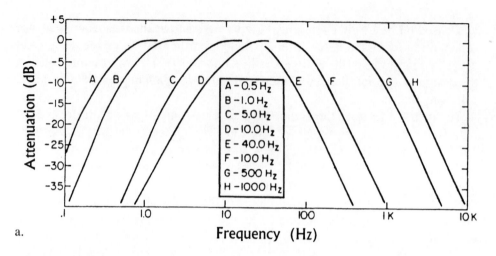

a.

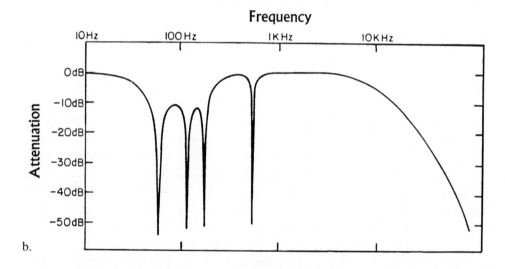

b.

Figure 6.7. Characteristics of filters designed for use in Penn State Mark II AE/MS monitoring system (after Kimble, 1983). a. Band-pass filter characteristics. b. Notch filter characteristics.

newsletter announcing the availability of their 3000 system, model 420 spatial discriminatory module. Figure 6.8 illustrates the associated transducer geometry. A similar arrangement could be used for three-dimensional situations. In Figure 6.8a, the "data transducer" is separated from a noise source, S, (e.g., mechanical grips, friction, testing machine noise) occurring to the right of the figure, by a so-called "guard transducer". In operation the noise signal will reach the guard transducer first, this will open a gate in the data monitoring circuit for a selected period of time preventing the noise signal from being recorded. In effect, a boundary AB is set up between the guard and the data transducers and only signals occurring to the left of this boundary (acceptance region) will be detected by the data transducer. Figure 6.8b illustrates a two-dimensional case, where your guard transducers are employed to establish a square acceptance region around the data transducer, with signals (S) occurring outside this region being rejected.

Spatial discrimination has been used extensively in large-scale industrial field tests and in laboratory tests where it is desired to eliminate noise generated by the testing facilities themselves. The application of spatial discrimination to the development of a low noise facility for studying AE/MS activity in geologic materials at very low stresses has been described in a paper by Hardy et al. (1989). Figure 6.9a illustrates the form of the specimen loading arrangement utilized. Here the test specimen was located between spherical upper and fixed lower-loading heads. An upper telfon cap and a lower telfon pillow provided acoustic isolation from the associated testing machine platens. Hardware-based electronic discrimination methods were utilized in an effort to further reduce the effects of noise generated in the specimen loading system. As shown in Figure 6.9b, AE/MS data were monitored both from a "data" transducer (T) mounted in the central region of the test specimen and from two "guard" transducers (G1 and G2) located at either end of the test specimen. In these studies the AE/MS monitoring systems, incorporated a Dunegan Model 420 spatial discriminator (SD), and provided three parametric outputs (specimen data, spatially discriminated (SD) data, and guard data). Normally these outputs

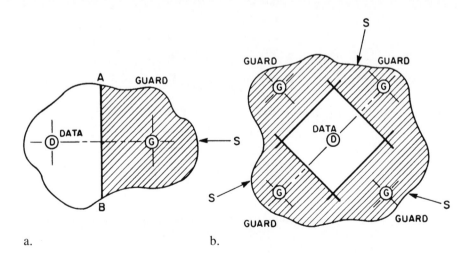

a.   b.

Figure 6.8. Illustration of spatial discrimination in one- and two-dimensions (after Pollock, 1979). (D-Data transducer, G-Guard transducer, S-Incoming noise signal). a. One-dimensional case. b. Two-dimensional case.

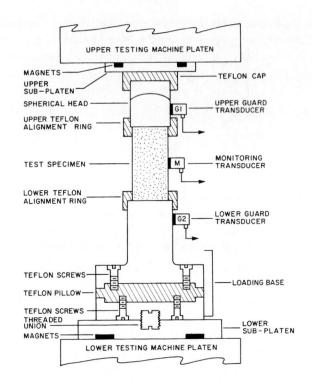

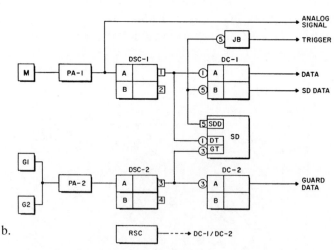

Figure 6.9. Low noise system used for monitoring of stress/strain thresholds for AE/MS activity in rock (after Hardy et al, 1989). (M and G1/G2 denotes AE/MS monitoring and guard transducers, respectively; PA, DSC, DC and RSC denote preamplifier, dual signal conditioner, dual counter and reset clock, respectively; SD denotes spatial discriminator; and JB denotes junction box. Square and round terminals denote output and input to the Dunegan 3000 system internal databus.) a. Specimen loading arrangement. b. Hardware-based spatial discrimination system.

were in the form of total counts, however, when the reset clock (RSC) was connected to the appropriate dual counter channel, equivalent count rate data was generated. The system also provided outputs (analog signal and trigger) to enable AE/MS analog signals associated with spatially discriminated events to be recorded on a Nicolet model 4094 digital oscilloscope.

Using the system described, the observed changes in strain thresholds for AE/MS activity observed in studies carried out on Indiana Limestone, Berea Sandstone and Tennessee Sandstone were found to be 3.8 µs, 2.0 µs and 1.0 µs, respectively. Such relatively low values add support to recent suggestions that so–called "secondary AE/MS activity" may occur in a variety of mechanical situations ranging from those associated with the evaluation of underground mine stability, to those involved in the development of meaningful earthquake prediction procedures. Additional laboratory studies on a greater variety of materials, and subsequent field studies are required to further substantiate the preliminary results obtained.

With the increasing use of multi-channel, computer-based AE/MS monitoring systems, a variety of additional spatial discrimination techniques are being utilized. For example, in geotechnical field studies zonal and more detailed source location techniques are being employed to define acceptance/rejection zones. In the latter case, the location and geometry of these zones may be easily modified by changes in computer keyboard input.

6.3.7   Special considerations

Monitoring facility optimization also depends on a number of special factors inherent in related electronic components of the associated monitoring system. In particular, consideration will be given here relative to the importance of the dynamic range, the Nyquist frequency, and analog signal reconstruction.

Dynamic range: Besides the various factors that establish the signal-to-noise ratio (SNR) of an AE/MS monitoring facility, discussed earlier in this section, an important aspect of a suitably optimized facility is the overall dynamic range of the monitoring system. According to Doebelin (1975) the dynamic range (DR) is the ratio of the largest (E_ℓ) to the smallest (E_s) dynamic input that an instrument will faithfully measure. It is normally given in dB, where

$$DR \text{ (dB)} = 20 \log_{10} (E_l/E_s) \qquad \text{(Eq. 6.1)}$$

For instance, if an instrument can handle input levels over the range of 1 to 1000, then DR = $20 \log_{10} (1000/1) = 20 \times 3 = 60$ dB. For example, a modified video recorder used for recording AE/MS signals, described earlier in Section 4.4.4, has a dynamic range of 26 dB. This indicates that it is capable of recording input levels ranging from 50 to 1000. Depending on the allowable maximum input signal, it may be necessary to use an attenuator to match the voltage of the output signal from the conditioning unit to the video recorder input.

Computer-based analog-digital (A/D) input facilities also have a specific dynamic range. Generally it is equivalent to 6 dB/bit. For example, an 8 bit A/D facility will have a DR = 48 dB, and is thus capable of processing input levels ranging from 4 to 1000. For example, if AE/MS data collected by a monitoring facility was to be analyzed for event amplitude distribution the computer-based facility could provide a much more detailed analysis than the video recorder system.

Nyquist frequency: When recording dynamic signals using digital systems great care must be taken to ensure that the observed frequency content of the signals are not influenced by the rate of signal digitization. For example, McConnell (1995) indicates that the maximum frequency of the input signal, (W_{max}), should not exceed one-half of the digital sampling frequency (W_s), namely:

$$W_{max} < W_s/2 \qquad \text{(Eq. 6.2)}$$

where W_{max} is known as the Nyquist frequency. This is normally accomplished by passing the signal through an anti-aliasing filter before it is sampled by the A/D converter. This insures that any high frequency components above W_{max} are significantly removed. The ideal is to be down 80 dB (for 16 bit A/D converters) at the break frequency (W_b), i.e. the 3 dB point. This means that the filter is down about 40 dB at the Nyquist frequency. For a filter with a fall-off rate of 120 dB/octave the break frequency is about 0.4 W_s, so that signal frequencies up to 40% of the sampling frequency may be used. For example, a 1 MHz sample rate allows one to analyze signals up to 400 kHz.

Analog reconstruction: When recording analog AE/MS data using a digital system the spectral quality of the recorded data depends on the Nyquist frequency. In contrast, in order to collect sufficient data to later reconstruct the original analog signal requires it to be sampled at least four or five times during the A/D process. For example, a minimum sample rate of 10 MHz is required to record and reconstruct 2MHz signals. Obviously the fine features of the reconstructed signals will improve with increasing sampling rate. However, the maximum sampling rate will be limited by the digital storage capacity of the system.

6.4   MONITORING FACILITY CALIBRATION

6.4.1   General

In essence, the term calibration refers to the procedure by which the output and input of a system are compared when the system input is varied in a defined manner using some form of standard (calibration unit). In the context of AE/MS studies the "system" could refer to an overall AE/MS monitoring facility, the various electronic components within the monitoring system, or the AE/MS transducer itself. Detailed literature exists relative to the general concepts of measurement system calibration (e.g. Cerni and Foster, 1962; Doebelin, 1975). In the AE/MS area material is available relative to the calibration of associated electronic components (Anon., 1996a) and transducers [sensors] (e.g. Anon., 1996b, Anon., 1996c, Miller, 1987). Various aspects of the latter have also been considered earlier in this text.

The current section will be limited for the most part to so-called "end-to-end" calibration, where some form of simulated AE/MS source is generated at a known location and compared with the signal obtained at the monitoring system output. Such calibration in effect examines the behavior of the complete monitoring facility, including transducer characteristics and coupling, the behavior of all electronic components in the monitoring system, associated signal processing systems, and readout facilities (e.g. printer, monitor, etc.) and all associated cables, connectors, junction boxes, and power sources. Thus, the term "end-to-end" calibration is appropriate.

Figure 6.10 shows a simplified block diagram of a generalized, one-channel AE/MS monitoring facility. It consists of a test structure (ST), an attached transducer (T), a monitoring system consisting of a preamplifier (PA), signal conditioning and data storage (SC/DS), and a readout unit (RO). A short cable (C1), and a long cable (C2) connect the major sections of the facility. The numbered triangular symbols indicate locations where signals may be applied for facility calibration . Calibration at location 1 involves the use of a mechanical type source, whereas at the other locations (2, 3 and 4) electrical-type sources are utilized.

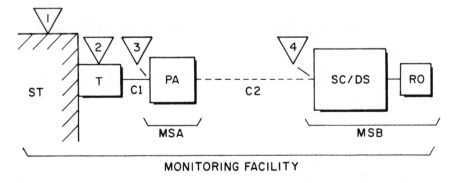

Figure 6.10. Block diagram of a simplified one-dimensional AE/MS monitoring facility. (ST – structure, T – transducer, PA – preamplifier, SC/DS – signal conditioning and data storage, RO – readout, C1 & C2 – short and long cables, and MSA and MSB – monitoring system components. Numbered triangular symbols indicate possible calibration locations.)

6.4.2 Laboratory calibration

Referring to Figure 6.10, calibration of laboratory monitoring facilities may involve application of electrical calibration signals at points 3 and 4, or mechanical signals at point 1. The latter provides the only true end-to-end calibration of the facility. Electrical calibration involves the injection of suitable electrical signals and the observation of effect of these signals at the output (RO) of the facility. A variety of conventional signal and pulse generators may be used to study such factors as the overall frequency response of the monitoring system, the amplitude linearity, event rate limits, etc. AE/MS signals are typically randomly occurring decaying sinusoids, rather than continuous sinusoidal or pulse-like signals. As a result, the so-called acoustic emission simulator (AES) developed by Acoustic Emission Associates is an extremely useful calibration tool. A photograph of this unit is shown in Figure 6.11.

The AES is capable of generating two separate channels of AE/MS-like signals for which various characteristics, e.g. repetition rate, waveform rise time, decay time and amplitude, may be independently selected. The characteristics of the two channels may be selected to be very different, and the two channels may be combined to provide a more complex calibration signal. Using the built-in delay capability (microseconds to seconds) of one or more dual-channel AE simulators, a program may be established for testing the source location performance of multi-channel AE/MS systems.

True end-to-end calibration of laboratory monitoring facilities is accomplished by the application of some form of mechanical calibration signal to the structure itself (Fig. 6.10, point 1). A variety of mechanical calibration techniques have been utilized. These include a pulse-driven piezoelectric

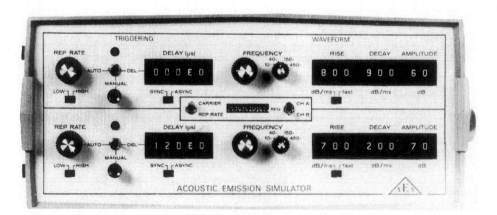

Figure 6.11. Acoustic emission simulator (AES) for electrical calibration of AE/MS monitoring systems.

transducer, a capacitative transducer source, a helium jet, pulsed lasers, a glass capillary break, ball impact, a spark source, an explosive source and the "pencil lead break" technique. A number of these are discussed in papers by Breckenridge et al. (1990) and Hsu et al. (1990). It should be noted such mechanical techniques are also useful for the evaluation of the propagation velocity characteristics of the test structure, and with multichannel monitoring systems, the evaluation of source location accuracy.

Probably the most common mechanical source is the pencil lead break (PLB) technique. The technique was originally devised by Hsu in 1975 (Hsu et al., 1977), although development this technique appears to have been underway by Nielsen (Anon., 1981) at about the same time. In this technique a length of lead projecting from the end of a mechanical pencil, is forced against the structure to be calibrated until it fails. This provides a mechanical point release force of the order of a few Newtons which depends on the size and type of lead used. Figure 6.12 shows the experimental arrangement. Traditionally, a Pentel pencil, with a length of 2H, 0.5 mm diameter lead projecting outward 3 mm, has been used. In order to ensure a constant breaking angle and reduction of spurious signals, a pencil mounted teflon guide ring and an associated recommended application procedure was developed (see Anon.,1981). Normally the calibration (lead breaking) process should be repeated a number of times. This reduces the standard deviation to approximately 5%. Generally the PLB technique is a convenient, inexpensive and relatively accurate method for end-to-end calibration of facilities for testing of laboratory specimens and laboratory scale structures. It has also been found to be useful in a variety of AE/MS field studies where relatively low attenuation materials are involved, e.g., metallic water, oil and gas lines.

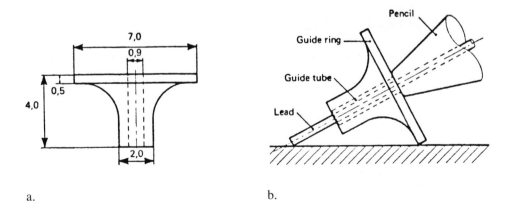

a.   b.

Figure 6.12. Arrangement for pencil lead break mechanical calibration (after Anon, 1981). (Dimensions shown are in mm.) a. Teflon guide ring. b. Calibration arrangement.

6.4.3 Field calibration

With the exception of metal structures like pipelines, storage tanks, bridges, etc., most field structures, particularly geotechnical ones, involve materials which are highly attenuative. As a result the mechanical calibration techniques discussed in the previous section are generally unsuitable. As in laboratory situations, electrical techniques may be employed to calibrate the monitoring system itself. In geotechnical structures "end-to-end" calibration normally involves the use of impact or explosive sources with energy output orders of magnitude larger than those utilized in laboratory applications. Table 6.1 lists a number of the more common sources which have been used for end-to-end calibration, velocity measurement, and source location accuracy evaluation in the field. Some of these have been discussed briefly earlier in the text.

Natural sources: A number of mechanical sources are active during mining, tunneling, and other geotechnical operations. These include activity generated by the excavation machinery itself (e.g. drills, and mining and tunneling machines) and adjustment of related geologic structures. In the former case, the use of mechanical sources generated by the excavation machines themselves have not had any major success. However, studies have shown

Table 6.1.   Mechanical sources for field applications.

Natural Sources*

- Excavation Machines
- Mining & Excavation Operations
- Seismic Activity

Impact Sources*

- Schmidt Hammer
- Down-Hole Hammer
- Near-Surface Hammer
- Surface Weight Drop

Explosive Sources*

- "Moose Gun"
- Sparker
- Gas Gun
- Water Gun
- Explosive Charges

---

*Source energy generally increases for each subsequent listing

(Dresen and Ruter, 1994) that shear waves (SH-waves) often accompany coal mining operations, and these in turn may excite typical Love seam waves. A paper by Frantti (1977) discusses the presence of a number of bands of low frequency seismic signals (9-10 Hz, 30-40 Hz and 80-150 Hz) apparently associated with structural adjustments of a tabular underground copper deposit due to mining. The review of the literature indicates a number of other similar situations.

Local as well as distant natural seismic activity may provide a suitable source for monitoring facility calibration. For example, Hardy and Mowrey (1981) report the collection of local as well as world wide seismic data using a high sensitive AE/MS monitoring facility capable of detecting signals with frequencies content as low as 1 Hz. In some cases nearby natural sources such as aircraft (Hardy et al., 1986), highway traffic (Hardy et al., 1988), and industrial operations may serve as useful mechanical sources.

Impact sources: A variety of impact sources have been used for AE/MS system calibration. These include devices such as the hand-held Schmidt hammer which has been found useful for surface impact of strong rock strata, (Hardy, 1974), and the so-called "down-hole hammer source" (Dresen and Ruter, 1994). The latter may be installed in a borehole at any desired depth and activated remotely. Surface and near-surface impact may be applied using simple facilities such as those shown in Figure 6.13. Here a hand held hammer, with weight up to perhaps 4 or 5 kg, is used to impact the ground surface, or using a suitable extension, impact the bottom-end of a borehole up to 1 m or so in depth. In certain cases the concept could be applied underground where impacts might be applied to the floor, roof or sidewalks of the associated opening. Depending on the depth and geometry of the actual point of impact, the system may generate P-waves, S-waves or a combination of both. Using the geometry illustrated in Figure 6.13, studies by Hardy et al. (1986) have shown that in soil, AE/MS signals with frequencies up to 50 Hz are generated with sufficient energy to propagate distances of 60 m or more. Larger propagation distances are possible with the use of larger impact energy. For example, as shown in Figure 6.14, the impact of masses varying from 20 to 50 kg were utilized by Hardy et al. (1986) in highway slope studies.

During field studies associated with AE/MS monitoring of a gas storage reservoir site (Hardy et al. 1981) mechanical calibration was carried out using a surface-applied impact source. This involved dropping an impact mass, with a weight of some 50 kg, from a height of approximately 3 m onto the ground surface above the transducer installation being evaluated. During initial tests the impact mass was allowed to contact the ground surface directly. However, this procedure was later modified and a thick steel impact plate (diameter–0.35 m and thickness–0.05 m) was located on the ground surface at the impact point. With this modification, it was found that more seismic energy could be transmitted to the soil and the underlying rock. A tripod was used to lift

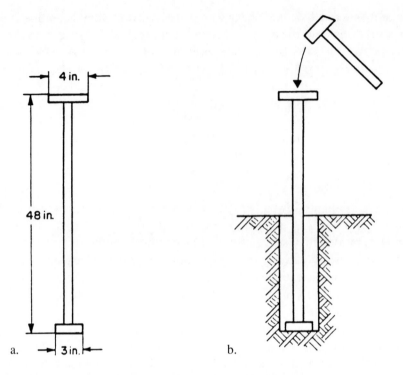

Figure 6.13. Near-surface impact technique (after Hardy et al. 1986). (Conversion factor: 1 in. = 2.54 cm.). a. Extension unit. b. Impact application.

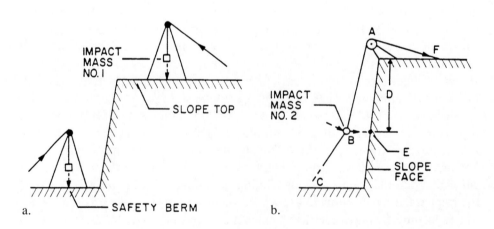

Figure 6.14. Two types of impact used for calibration studies on rock slopes (after Hardy, et al. 1988). (Mass no. 1 »50Kg, Mass No. 2 ≈ 20 kg.) a. Slope top and berm tests. b. Slope face tests.

the impact mass to the desired height, and a geophone to provide the time of impact, was close to the impact point. It should be noted that the seismic signal generated was clearly received by a nearby shallow burial transducer (PSU-2S), by a transducer (PSU-2D) located at a depth of approximately 150 m directly below the impact point, as well as by other down-hole transducers as far away as 600 m. In most cases, the detected calibration signal showed both P- and S-wave components. However, the signal associated with the transducer directly below the impact point (PSU-2D) showed only an S-wave component. Finally, Nagashima et al. (1992) describe a unique in-situ calibration procedure which involved the use of seismic signals generated by an operating percussive drilling system.

Explosive sources: A variety of explosive devices have been used as seismic sources (Dresen and Ruter 1994) including the "moose gun," the "sparker," gas and water guns and explosive charges. A number of these may be used for monitoring facilities calibration. The term "moose gun" was coined by geophysicists working in areas of northern Canada where herds of moose were often present. Basically this technique utilizes shotgun shells which are detonated in shallow, water or mud filled, holes. In its most primitive form the shotgun shell is locked in place at the end of a section of steel pipe, perhaps a meter or so in length, the steel pipe is placed vertically in the calibration borehole, and a steel rod is dropped into the open end of the pipe where it detonates the shotgun shell. The resulting energy is transmitted outward through the surrounding water or mud and couples into the surrounding rock or soil. If required, a nearby geophone may be used as a means of determining the time of source initiation. Calibration studies, using this type of simple source, have been carried out by Hardy et al. (1986). Studies using this type of source in deep boreholes are described by Dresen and Ruter (1994). The down-hole tool developed for this purpose is shown in Figure 6.15. It has a diameter of 6.4 cm, a length of 35.6 cm, and a weigh about 4.5 kg. A neoprene diaphram near the end of the barrel keeps borehole fluid out of the firing chamber. The firing sequence is initiated by a switch closure keyed to the monitoring system recorder, activating the solenoid, releasing the firing pin mechanism, and subsequently detonating the shotgun shell. The resulting closure of a built-in inertia switch on the tool provides the zero-time pulse on the recording. To date this device has been operated at depths up to 150 m.

A wide range of larger explosive charges have been used as sources for AE/MS facility calibration. As noted earlier, routine and maintenance related blasts at mining and other geotechnical sites can often be utilized for calibration purposes. For example, the author has used blasts ranging in size from a single stick of dynamite to thousands of kilograms of explosives, such as ANFO and TOVAN, as calibration sources. In the case of large blasts a series of delays are involved in the blasting sequence with individually detectable blasts ranging from 45 to 180 kg/delay. Hardy and Beck (1977) describe the use of small

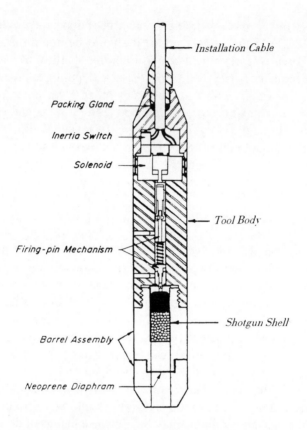

Figure 6.15. Down-hole explosive source utilizing shotgun shells (after Hasbrouck and Hadsell, 1976; see also Dresen and Ruter, 1994).

explosive charges for the development of velocity models at a underground coal mine site. Here 1/2 to 2 sticks of 40% ammonia gelatin dynamite were used along with standard electric blasting caps. In these studies it was found that special seismic blasting caps, for accurate monitoring of detonation time, were too expensive and special elect ronic circuits were developed for this purpose. Dresen and Ruter (1994) discuss the use of other types of explosive sources, including a special SH-wave source consisting of a length of Dynacord formed into a spiral-shape and detonated in a suitable borehole.

6.4.4   In-situ geophone calibration

Certain types of high quality research-type geophones may be purchased with built-in calibration capabilities, see calibration location 2, shown earlier in Figure 6.10. An example of this capability is the HS-10-1/B, dual-coil very-low-frequency geophones (seismometers) used by Hardy et al. (1981) for

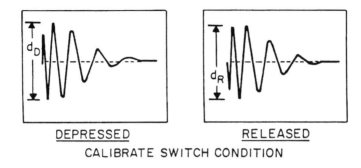

DEPRESSED   RELEASED
CALIBRATE SWITCH CONDITION

Figure 6.16. Typical waveforms generated during insitu geophone calibration, and associated peak-to-peak deflections d_D and d_R.

AE/MS monitoring at the New Haven underground gas storage reservoir site. These transducers may be calibrated electrically through the use of a separate built-in calibration coil operated from a remote position . In use, a specified level of current is applied to the calibration coil, via an extra set of wires in the geophone cable. This current is held for 10 seconds, causing the geophone sensing coil to be deflected in a specific manner. The applied calibration coil current is then removed, causing the calibration coil to be deflected again. The resulting calibration waveforms are shown in Figure 6.16.

The peak-to-peak chart deflections, d_D and d_R, shown in Figure 6.16, provide the required calibration data. To allow for possible asymmetry in the system the average deflection is utilized, namely:

$$\bar{d} = (d_D + d_R)/2 \qquad \text{(Eq. 6.3)}$$

where $\bar{d}$ is the calculated average peak-to-peak chart deflection, and d_D and d_R are the measured peak-to-peak deflections associated with the depression and subsequent release of the calibration switch. The peak-to-peak particle velocity for each transducer, associated with the standard electrical calibration, is computed from the calculated average peak-to-peak deflection, the overall monitoring system gain, and the computed transducer sensitivities. Transducer sensitivities are computed using an equation provided by the manufacturer. For example, Appendix III in Hardy et al. (1981) provides the required equation (Equation III-2) for the HS-10-1/B geophones used at the New Haven field site. In-situ geophone calibration at this site was carried out on all six permanently installed geophones some 14 times over a 1000-day test period. Equivalent calibration values were of the order of 50.8 x 10^{-6}m/sec, with percent standard deviations ranging from 6.7% to 14.0% depending on the particular transducer.

6.5   VELOCITY MEASUREMENT

6.5.1   Introduction

As indicated earlier, in Section 5.8 (Velocity Models), accurate AE/MS source location generally necessitates high quality velocity for the specimen or structure under study. Generally the determination of velocity data for laboratory specimens and small structures is relatively straight-forward. Here, the necessary tests may be carried out using various acoustic sources, such as pencil lead breaks, piezoelectric elements, helium jets, etc., and one or more suitably located receiving transducers. For example Rothman (1977) describes the use of piezoelectric transmitting and receiving crystals to determine the 3-dimensional velocity model for sandstone test specimens during creep experiments.

In contrast, geotechnical field projects involve structures which are usually large, and dimensions may vary from hundreds to thousands of meters. As a result, in acoustic emission/microseismic (AE/MS) studies associated with such structures, the ability to determine accurate values of the source location coordinates is of primary importance. A variety of factors affect source location accuracy. However, past experience has indicated that one of the most critical factors is wave propagation velocity. Furthermore, in most field situations it is difficult, if not impossible, to assume that the velocity is uniform and isotropic. As a result, some form of velocity model is necessary to accurately describe the velocity characteristics of the structure under investigation. A variety of velocity models have been investigated for use in field-scale studies carried out in geologic materials. These include isotropic, anisotropic and "unique" half-space models, and a multi-layer isotropic model. Hardy (1986) has reviewed this topic in some detail and a brief outline is included here.

6.5.2   Field techniques

In essence, the fundamental concepts involved in acquiring seismic velocity data for a material in the field, are to locate a suitable seismic source and detectors (transducers) at known positions in the material and to monitor the travel time of seismic signals propagating from the source to the detectors. Knowing the distances between the source and the detectors, the source initiation time (T_0) and the time of arrival of the seismic signals at the detectors it is possible to compute the associated seismic velocities.

As indicated earlier in Section 6.3 (monitoring system calibration), a variety of seismic sources are available for AE/MS related field studies. These include, in approximate order of decreasing energy, explosive charges, mechanical impact, airgun, "sparkers" and a variety of sources incorporating magnetostrictive and piezoelectric elements. Depending on the assumed velocity complexity of the structure, or the desired sophistication of the developed velocity model, one or a number of detectors (transducers) would be required. Furthermore, the selected sources and detectors must be capable of generating and detecting the appropriate type of seismic signal over the desired frequency range.

It should be noted that a variety of types of seismic signals, and associated seismic velocities, are possible (e.g. P-wave, S-wave, Rayleigh wave, etc.). Furthermore, it is important to realize that the seismic velocities associated with even a relatively uniform geologic structure are intrinsically anisotropic and often affected by a variety of other factors, such as in-situ stress, temperature, presence of fluids, etc. As a result, some form of velocity model is necessary to describe the velocity characteristics of a specific site. It should be noted, however, that the development of an overly complex velocity model for a specific structure is generally unrealistic, and relatively simplified models must normally be utilized.

A variety of techniques have been utilized to acquire the data necessary for development of a meaningful field velocity model. For convenience these techniques will be subdivided as follows: (1) Seismic Transmission Techniques, (2) Seismic Refraction Techniques and (3) Borehole Logging Techniques.

To date, the majority of AE/MS field studies have been associated with subsurface activities, such as mining, tunneling and underground construction. Here, seismic transmission techniques have been used almost exclusively for acquisition of velocity data. in recent years, however, a variety of somewhat unique areas of AE/MS investigations have been underway. Here seismic transmission techniques are less applicable and techniques involving surface seismic refraction and borehole logging techniques have been investigated. In general, past experience has indicated that velocity data obtained from laboratory-scale tests are not suitable for the development of meaningful field scale velocity models. Nevertheless in the absence of any other data, data obtained using such tests may be of at least limited usefulness.

6.5.3 Seismic transmission techniques

Seismic transmission techniques represent the most direct means of obtaining seismic velocity data. This may involve either an in-place AE/MS monitoring system or a system specifically installed to acquire velocity data. Figure 6.17 illustrates three different types of applications where seismic transmission techniques are employed, namely: subsurface, surface-subsurface and near-surface.

Subsurface applications: Figure 6.17a, illustrates two subsurface applications, namely the acquisition of in-seam and overburden velocity. For the most part, acquisition of seismic velocity data at depth using seismic transmission techniques involve the use of explosive sources. In a limited number of cases,

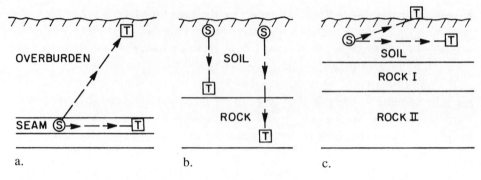

Figure 6.17. Various applications where seismic transmission techniques are used to acquire velocity data. (Symbols S and T denote seismic source and transducer (detector), respectively.) a. Subsurface. b. Surface-subsurface. c. Near-surface.

mechanical impact type sources have also been utilized. Generally subsurface explosive sources may be classified as follows:
1. Industrial Sources
   • Scheduled, Production-related Blasts
   • Random, Maintenance-related Blasts
2. Research Sources
   • Large-scale Blasts
   • Small-scale Blasts
   • Mechanical Impact

In many mining and excavation related projects, scheduled, production-related blasts are commonly used to establish and update the in-situ velocity data necessary for AE/MS source location. Blast location coordinates are normally available, although they may be somewhat less accurate than desired. In most cases, source time (T_0) is not available and data from a number of transducers are required to evaluate travel times and associated velocities. Depending on the situation, scheduled production-related blasts may involve charges ranging in size from 4 to 4000 kg or more. Due to safety restrictions, production-related blasting is not normally carried out in coal mining operations. However, in most mining and excavation related projects, small randomly occurring blasts (e.g., 1/2- 2 sticks of dynamite) are associated with various maintenance activities. In the majority of cases, however, neither the blast coordinates or the associated T_0 data are available for these sources. As a result, except under special circumstances, such data are generally unsuitable for velocity model development.

Research sources used in subsurface situations may involve dedicated blasts, ranging from large-scale (> 40 kg) to small-scale (< 4 kg), or mechanical impacts generated by a hand-held sledge hammer, Schmidt hammer, or similar device. The type of source will depend on the dimensions of the structure under evaluation. In order to provide the most suitable data for velocity computation, T_o and source coordinates are required. T_o data may be generated by the use of a suitably located transducer, or in the case of explosive charges, by special blasting caps or blast detection circuits (see Hardy and Beck, 1977).

Surface-to-subsurface application: In a number of field situations direct access to the underground structures for which velocity data is required may not be possible, or restrictions on the underground use of the required type or magnitude of seismic source may exist. In such cases one approach is to locate the seismic source, and an associated T_o transducer, on the surface. Downhole transducers, or transducers located in the structure under study, are then utilized for detecting the arrival of the transmitted seismic signal. Figure 6.17b, presented earlier, illustrates two such applications. Sources are normally explosive charges or mechanical impact, however, for shallow depths (d < 30 m) and competent material, other types of sources (e.g. air gun) may also be suitable.

Figure 6.18 illustrates the application of this technique for the evaluation of the P-wave velocity of the regolith overlying an underground gas storage site (Hardy, et. al, 1981). Here a 45 kg impact mass was dropped from a height of a few meters on to the ground surface some 150 m vertically above a permanently installed downhole transducer (PSU-2D). This transducer provided arrival time data and a permanently installed, near-surface transducer (PSU-2S) was used to provide T_o data.

A similar approach was used by Hardy and Beck (1977) to evaluate the average P-wave velocity of the rock strata overlying a longwall coal mine in Central Pennsylvania. Here, access to the mine was available and suitable transducers were mounted to a thick steel plate attached to the mine roof using steel bolts and epoxy. Seismic sources consisting of 1/2 to 1-1/2 sticks of 40% ammonia gelatine dynamite were detonated on surface some 150 m vertically above the underground transducer. T_o values were determined using a special electronic circuit activated directly by the explosive charge. The depth of soil at the detonation site, necessary to compute true rock velocity, was determined using seismic refraction methods.

Near-surface applications: Seismic transmission techniques are also applicable to many near-surface situations. Here both seismic sources and detectors may be located directly on surface or in shallow boreholes. Borehole depth will depend on the location of the soil or rock mass to be investigated. Particular care must be taken to insure that directly transmitted, rather than refracted, signals are utilized in the associated analysis.

Due to weathering, high seismic attenuation is evident near surface, and both velocity and attenuation are found to vary considerably with position. Furthermore, in near-surface applications of AE/MS techniques, the scale of the structure under study may be considerably smaller than that encountered

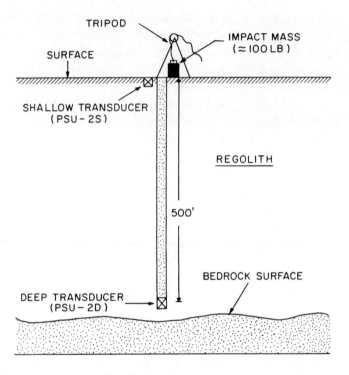

Figure 6.18. Experimental arrangement used to determine seismic velocity in the regolith layer overlying an underground gas storage site (after Hardy et al., 1981). (Conversion factors: $1' = 0.3048$m, 1 lb = 0.4536 kg.)

in the subsurface. As a result, source to detector distances are often relatively small (e.g. 15-30 m), and more detailed velocity studies, such as those based on tomographic techniques, may be required. Recent AE/MS studies relative to the stability monitoring of sinkholes at a Pennsylvania airport (Hardy et al., 1986) discuss the procedures for acquiring seismic velocity data at a near-surface site. Here mechanical impact (sledge hammer), and small explosives sources (blank shotgun shells) have been used successfully as seismic sources.

6.5.4   Seismic refraction techniques

In some field cases, access to the structure under study for the purpose of installing seismic sources and detectors is impossible. In such cases, seismic refraction techniques (Dobrin, 1960), which employ both near-surface sources and detectors, may be utilized. Figure 6.19 illustrates a simplified experimental arrangement for seismic refraction studies. Here the seismic signals from the source S travel, both directly and along specific refracted paths, to the various

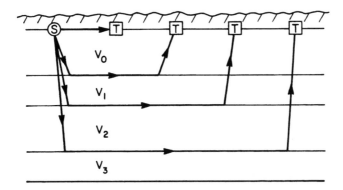

Figure 6.19. Simplified experimental arrangement for seismic refraction studies. (S and T denote seismic source and transducers (detectors) respectively. V_0, V_1, V_2, & V_3 denote the velocity of the various rock layers, where here it is assumed that $V_0 < V_1 < V_2 < V_3$.)

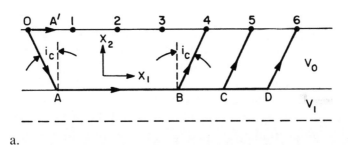

a.

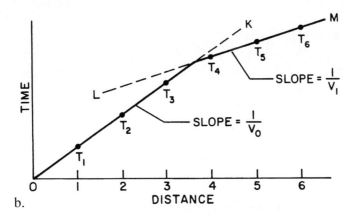

b.

Figure 6.20. Seismic refraction concepts (after Dobrin, 1960). (In (A), 0 denotes seismic source, 1, 2, ..., 6 denote transducers (detectors). In (B),T_1, T_2, ... ,T_6 are the arrival times at the various transducers. It is assumed that $V_0 < V_1$.) a. Ray paths of least time. b. Time-distance curve.

transducers which are arranged in a linear array. The path and the associated arrival times depend on the seismic velocities of the various layers.

Figure 6.20a illustrates the refraction concept in more detail in terms of a two-layer structure. Here initially the velocity of the two layers are considered to be isotropic and to have velocity values of V_0 and V_1, where $V_1 > V_0$. Seismic energy released by the source at point 0 will travel radially outwards. In particular, one component will travel by a direct path 0A′ towards the transducers located at points 1, 2, ..., 6, and a second will travel along the refracted path 0AB. When the ray 0A strikes the interface between the two layers at the critical angle, $i_c = \sin^{-1}(V_0/V_1)$, the energy is transmitted horizontally along the interface between the two layers at a velocity V_1. This generates seismic disturbances in the upper layer which propagate spherically outwards at velocity V_0. Certain of these rays (e.g. B4) propagate at an angle i_c and reach the surface transducers before the associated direct path signal.

Figure 6.20b illustrates the time-distance curve for the two-layer model shown in Figure 6.20a. Here it is indicated that transducers 1, 2 and 3 receive the direct signal at times T_1, T_2 and T_3 before they receive the refracted signal. Transducers 4, 5, and 6 in contrast receive the refracted signal first at times T_4, T_5 and T_6. Plotting the data as shown, allows two straight lines OK and LM to be constructed, the slopes of which represent the reciprocals of the average velocities of the two rock layers. Similar techniques are available for structures with three and more layers.

If the velocity characteristics of the two layers were not isotropic (e.g. different in the x_1- and x_2-directions) it is assumed that the situation would be more complex but of similar form. Although value of i_c would differ from the isotropic case, once the energy began to move along the boundary between the two materials it would travel with a velocity $V_1 (x_1)$, equal to the horizontal component of the velocity field in the lower layer. It is also clear that since the direct path from the source to the various detectors is along the x_1 direction even though the velocity is anisotropic the computed values of velocity in the upper layer are always the horizontal component $V_0 (x_1)$ of the velocity field. In review then, it appears realistic to assume that the values of velocities determined using the seismic refraction technique represent the horizontal velocity components in a direction parallel to that of the linear array used to acquire the refraction data.

Seismic refraction techniques have been used by Hardy and Beck (1977) to evaluate the anisotropic velocity field associated with the strata overlying the Greenwich coal mine in central Pennsylvania. The coal seam at this mine averaged about 1 m in thickness and was overlain by an average cover of 130 m. Soil depths above the mine ranged from 3 to 6 m. Longwall panels were commonly 140 m in width and 1500 m in length. During the study, a total of 15 AE/MS transducers were installed in four parallel rows over a 300 m long section of a selected longwall panel (B-4). AE/MS activity was monitored as the mining face approached and passed under the installed transducer array.

The velocity models developed for the Greenwich site were based on data available from two techniques, namely: seismic transmission and seismic refraction. Initially, a simplified anisotropic half-space was developed. As shown in Figure 6.21a, refraction data was obtained along two perpendicular seismic lines oriented at S30W and N60W. Seismic transmission data were obtained in the vertical direction using a surface blast and an underground transducer (detector) mounted on the roof of the mine. The x-, y-, and z-components of the velocity, as shown in Figure 6.21a, are approximately 3000, 2910 and 3275 m/s, respectively.

From a geological point of view there is no reason to suspect that any velocity anisotropy should exist in the horizontal (X-Y) plane. The relatively small difference in the values of V_x and V_y (88 m/sec) suggest that it might be reasonable to assume a uniform average velocity V_h of 2955 m/sec in the horizontal plane, i.e., the horizontal cross-section of the ellipsoid in Figure 6.21b is a circle with a radius of 2955 m/sec. Unfortunately, source location analysis using the anisotropic half-space velocity model, based for the most part on seismic refraction studies, was found to yield unsatisfactory results. It should be pointed out that the field data utilized to develop the anisotropic

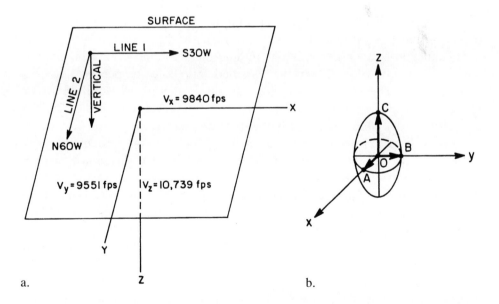

a. b.

Figure 6.21. Greenwich anisotropic half-space velocity model (after Hardy, 1986). (fps denotes ft/sec. Conversion factor 1 ft/sec = 0.3048 m/sec.) a. Data obtained from transmission and refraction studies. b. Velocity ellipsoid presentation of the anisotropic half-space model.

model was obtained at a location some 1200 m from the monitored longwall. Furthermore, as indicated earlier, due to a lack of sufficient field data the true magnitude of the horizontal velocity components (V_x and V_y) are uncertain. It is felt that a suitable anisotropic half-space model could have been developed, however, due to time constraints this was not possible. Instead, a so-called "unique velocity model" was developed based on seismic data generated by a small mining related blast occurring underground adjacent to the longwall face under study. The use of this model provided satisfactory source location results.

6.5.5   Borehole logging techniques

Geologic structures, where well-defined layering is suspected, and where access for the installation of a suitably refined array of transducers (detectors) and seismic sources is not possible, present special problems. One approach, as illustrated in Figure 6.22a, is the use of borehole logging techniques. Here one or more boreholes are drilled into the structure, or use is made of existing uncased boreholes. The selected borehole is first filled with drilling mud, then a suitable logging tool (sonde) is lowered down the length of the borehole to provide a graphic representation of the velocity characteristics of the adjacent geologic formations. In its simplest form, the sonde consists of a sonic transmitter and receiver separated by a known distance. Travel-time is recorded for a P-wave pulse traveling from the transmitter to the receiver. The inverse of this travel-time is the P-wave velocity of the formation.

Complications, such as variations in hole-diameter and in tool-inclination, have led to the development of the so-called "borehole-compensated" (BHC) system. The BHC sonde allows for the above factors by employing two transmitters and two separate pairs of receivers as shown in Figure 6.22b. The transmitters, one above and one below the receivers, emit pulses of acoustic energy. Each pulse travels through the drilling mud, into the formation several inches, and through the formation parallel to the hole. The difference between the time the pulse is detected by the first and the second receivers (i.e., R_2 & R_4 and R_3 and R_1) is divided by the distance between the receivers to provide a Δt value. The transmitters (upper and lower) are pulsed alternately, and the resulting Δt values are automatically averaged by a computer in the associated logging facility to give a compensated Δt value for the interval, usually expressed in μsec/ft.

In addition to the continuous travel-time curve, the typical sonic log also includes a record of hole diameter and an integrated travel-time scale. The hole diameter, provided by a caliper mounted on the sonic tool, is important to sonic log interpretations since the hole is commonly "washed out" immediately below the casing on a cased hole, resulting in a larger than normal hole diameter. This may result in the sonic log being unreliable in this region, due to an excessive

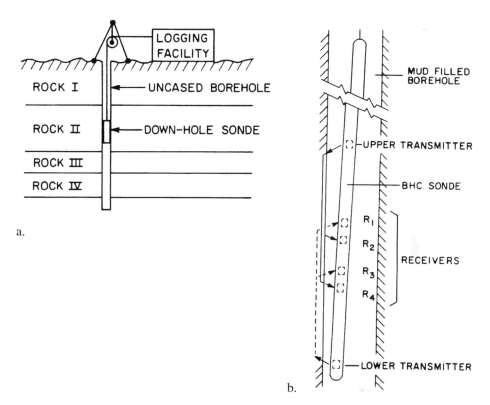

Figure 6.22. Outline of borehole logging concept and typical BHC sonde. a. Borehole logging concept. b. Schematic of typical BHC sonde (after Anon., 1969).

thickness of drilling mud in the sonic travel-path, causing anomalously high travel-times. The integrated travel-time information is displayed by a series of "pips" adjacent to the sonic curve. Each pip is separated from the previous one by an interval of one millisecond. The travel-time between two depths may be obtained by simply counting the pips. The integrated travel-time is useful for obtaining average sonic velocities in the formations surrounding the hole. Figure 6.23 illustrates a typical sonic log including a caliper log, a total-integrated travel-time curve, and a BHC sonic log. As indicated previously, the Δt value presented in the sonic log is the inverse of the sonic velocity for the interval.

In using this technique to obtain velocity data for AE/MS source location, consideration should be given to a number of factors. First, the velocity data obtained is only characteristic of the material located within a few inches of the borehole. Damage to this material due to drilling or localized high stresses could result in abnormal values of velocity. Second, the frequency of the seismic (sonic) signals employed are considerably higher than those associated with

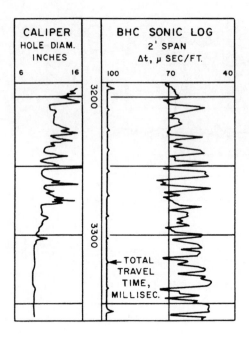

Figure 6.23. Typical sonic borehole log, including caliper log, total integrated-travel-time data, and BHC sonic log (after Anon., 1969).

typical AE/MS field signals (0.5 Hz to 1000 Hz) and the frequency dispersion (dependence of velocity on frequency) effects may have to be considered.

Borehole logging techniques were utilized by Hardy et al., (1981) to acquire seismic velocity data during AE/MS studies at an underground gas storage reservoir at New Haven, Michigan. The gas storage strata was located at a depth of approximately 300 m and was overlaid by a series of sedimentary beds and some 150 m of regolith (glacial tile). As a result a multi-layer velocity model was considered appropriate for source location at this site.

The basic data available for the development of this model were the regolith-velocity, 2134 m/sec (7000 fps), determined from seismic transmission studies and a set of borehole logging data. Figure 6.24 illustrates this data graphically as a function of position below surface. The velocity data was regrouped in a variety of ways in order to obtain the most suitable velocity model for use in source location analysis. This included regrouping in terms of the major geologic units, and various multi-layer models including nine-, five-, three-, two- and one-layer models. The quality of the various models was based on the level of standard deviation within each of the associated layers (i.,e. strata intervals). Models with a large number of layers (> 10) were not considered applicable since a layer less than half a wave length in thickness would probably have little effect on the propagation of a given elastic wave, unless it differed radically in physical properties from the immediately-adjacent layers. Since the shortest wavelengths (highest frequencies) being monitored were on the order 30 m to 150 m, a model of more than 10 layers was considered to be overly detailed. A review of the standard deviation information indicated that, although the grouping in terms of major geologic units was reasonably satisfactory, the nine-layer model shown in Figure 6.25 was a much better fit than any of the other groupings. Utilization of a model with fewer than nine layers probably represents a simplification of the actual situation, although such may be necessary for a mathematically-manageable source-location analysis. For example in the New Haven studies, the two-layer model shown in Figure 6.26 was found to be suitable in many cases.

6.6 WAVEGUIDES

6.6.1 Introduction

In the AE/MS area, a waveguide is defined as a component for conveying an AE/MS signal from a test specimen located in a hostile environment, or from a remote area of a large test structure, to a transducer located in a convenient benign environment. Such waveguides are normally composed of a low attenuation solid, such as steel, but in some cases may involve a

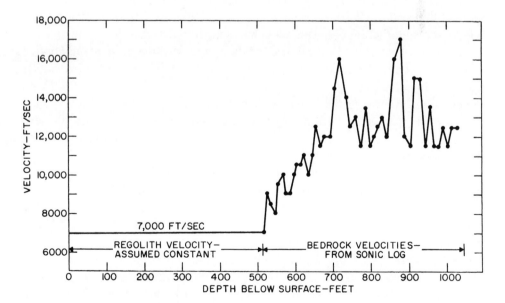

Figure 6.24. Variation of P-wave velocity data as a function of depth at New Haven underground gas storage site (after Hardy, et al., 1981). (Conversion factors: 1 ft = 0.3048m, 1 ft/sec = 0.3048m/sec.)

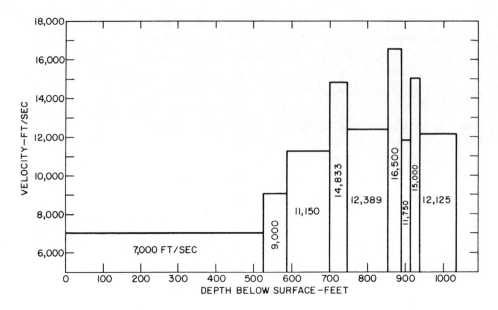

Figure 6.25. New Haven nine-layer velocity model (after Hardy et al., 1981). (Conversion factors: 1 ft = 0.3048m, 1 ft/sec = 0.3048m/sec.)

fluid component. The general topic of mechanical waveguides, relative to the propagation of acoustic and ultrasonic waves is considered in detail in a classic text by Redwood (1960). Current texts on acoustics and ultrasonic also deal with associated theoretical aspects such as guided wave theory.

A review of the AE/MS literature indicates a lack of theoretical and applied studies relative to the use of waveguides. There are frequent comments on the use of various forms of waveguides in field studies at hostile test sites (e.g. high temperature processing facilities, nuclear reactors, etc.), however, few details are given relative to the design, installation, and acoustic properties of these components. In particular, in most cases, little or no consideration is given to the effect of the waveguide on the characteristics of the AE/MS signal.

In the general literature, an early brief note by Dunegan (1982) considers the effect of waveguides on AE/MS monitoring including the types of wave modes transmitted and the problem of multiple reflections. Wood et al. (1990) considers the theoretical aspects of waveguides developed for use in AE/MS monitoring of a high temperature pressure vessel operating at 580°C, a large cryogenic storage tank operating at –40°C, and a fibre reinforced containment vessel operating at 96°C. A recent paper by Dunegan (1991) investigates the frequency characteristics of simulated AE/MS sources in a special test structure (damped bar) detected by transducers attached by direct contact and by different types of metallic waveguides. These studies clearly illustrate the effect of the waveguide on selective transmission of P, S, and surface waves,

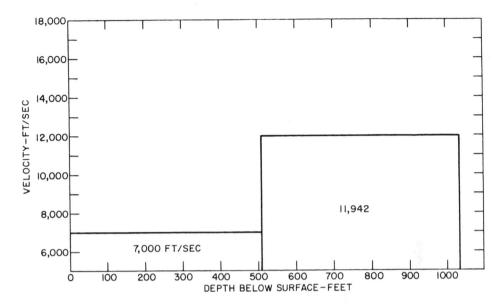

Figure 6.26. New Haven two-layer velocity model (after Hardy et. al., 1981). (Conversion factors: 1 ft = 0.3048m, 1 ft/sec = 0.3048m/sec.)

and the problems of multiple reflections. Zhang (1989) and others have utilized waveguides to monitor pipelines and other similar structures.

When using waveguides, it is important to consider their effect on the overall AE/MS monitoring system. For example, Figure 6.27 illustrates the application of a waveguide to the monitoring of a test structure (ST) located in a hostile environment (e.g. high temperature). In Figure 6.27a, a suitable waveguide is shown coupled to the test structure. Outside the hostile environment a transducer (T) is coupled to the free end of the waveguide. In operation acoustic signals (AS) in the structure travel down the waveguide, activate the attached transducer, and generate electrical signals (ES) which are further processed by an associated AE/MS monitoring system. Initial consideration suggests that this arrangement is very similar to direct attachment of the transducer to the test structure except that the waveguide provides a convenient "extension" of the structure outside the hostile zone. Figure 6.27b illustrates that, from a wave propagation point of view, the situation is in fact considerably more complicated, since effects due to waveguide coupling to the structure and the transducer (A_1 and A_2) must be considered, along with the effects of the waveguide itself.

The use of waveguides in field and laboratory studies on geologic materials has been briefly noted earlier in Sections 3.5.3 and 4.3.2 of this text. In the geotechnical area there is a critical need for high-fidelity waveguides due to remote location of critical monitoring sites and the high attenuation

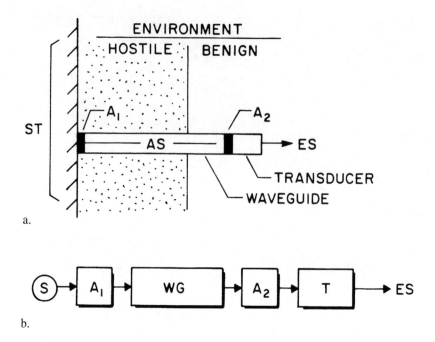

Figure 6.27. Waveguide arrangement and equivalent coupling system. (ST – test structure, S – AE/MS source, WG – waveguide, T – transducer, A_1 & A_2 –acoustic coupling, AS – acoustic signal, ES – equivalent electrical signal.) a. Mechanical arrangement. b. Equivalent coupling system.

characteristics of geologic materials. In this section the concept and application of waveguides will be discussed in further detail.

6.6.2   Waveguides for geotechnical studies

Although some of the material included in this section may also be applied to laboratory scale studies (e.g. test specimens, laboratory scale models, etc.) consideration for the most part, will be given to geotechnical field applications. In many of these applications high frequencies (> 20 kHz) are normally desirable either because the study is carried out in an area with high ambient noise and extensive low frequency filtering is required, or because only small regions of the structure have to be monitored. However, such frequencies are attenuated quickly due to the highly dissipative characteristics of the associate media, and to overcome this problem, mechanical waveguides have been utilized with the aim of increasing the zone of influence of the transducer. For example, Figure 6.28 illustrates the use of waveguides installed to monitor slope stability. The assumed principle of such a waveguide in this case is that, since it is normally made of a material with a much lower attenuation coefficient than that of the

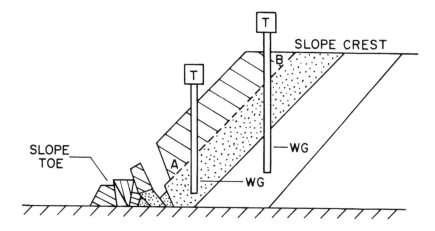

Figure 6.28. Waveguides installed to monitor slope stability. (AB – plane of weakness, WG – waveguide, T – transducer.)

structure (often soil), it should be a highly efficient method of coupling the transducer to the structure under study. Several researchers have reported very good results using waveguides. Despite the apparent success in some situations, the behavior of waveguides as an AE/MS transducer-structure coupling system has not been investigated in any detail, and its broad applicability has been questioned by some writers (e.g., Hardy and Mowrey, 1977).

Relatively few papers have been published on transducer-coupling systems for seismic and AE/MS geological-geotechnical applications. In the seismic area Washburn and Wiley (1941) studied the effect of seismometer installation on the motion imparted to the seismometer by the ground. They concluded that buried and wooden-post-mounted seismometers had better response characteristics then surface planted seismometers, however, their study was limited to an upper frequency of 500 Hz. Later Fail et al. (1962) obtained the same general results and predicted that high frequency applications (greater than 500 Hz) might be more complicated. Lamer (1970), based on a comprehensive theoretical study, concluded that, for the case of seismic signals in the petroleum field (i.e., frequencies lower than 500 Hz), the movement of the geophone coil is the same as the movement of the soil without the geophone, as long as the soil is homogeneous. A recent paper by Cranswich and Sembera (1988) describes a relatively detailed study to evaluate the effect of geophone site characteristics on recorded seismic data.

Koerner et al. (1977), in a comparative soil behavior study, evaluated some characteristics of waveguides used as a AE/MS coupling system. One of their conclusions was that the soil mass around the waveguides does reduce transducer sensitivity, but they did not quantify the level of such loss. Lord et al. (1982) evaluated the utilization of steel rods as AE/MS waveguides with the objective of determining the attenuation of the signal traveling in the steel rod due to energy absorption by the surrounding soil, which they defined as "soil covering loss". They concluded that, using typical values for the parameters, about 30 times more soil can be sensed using a waveguide than with the same transducer placed on the surface. Hardy (1985) considered the specific use of waveguides in AE/MS monitoring systems where source location data are required. He suggested that one solution to the problem is to attach a transducer to both ends of the waveguide. By determining the difference in arrival-time at the two transducers it is then possible to locate the point of AE/MS activation. Nakajima (1988) and Nakajima et al. (1995) have recently utilized the dual-transducer concept for AE/MS monitoring of landslides.

6.6.3   Geotechnical waveguide limitations

From wave propagation theory, it is known that a seismic wave propagating in a solid media is attenuated by several mechanisms (geometric spreading, internal friction, scattering and mode conversion), and that the frequency content of the wave is modified. Now consider the case of a soil structure which is being monitored using a perfectly flat-response transducer mounted on the upper-end of a waveguide. If a specific mechanical instability is sufficiently strong, the associated seismic wave will propagate through the material and reach the waveguide. Analysis of the data, detected and propagated by the waveguide, will be somewhat complicated. Firstly, since the material normally used for waveguides is steel, with an approximate propagation velocity of 5,000 m/s, and since the average soil velocity is of the order of 1,000 m/s, considerable mode conversion will occur due primarily to reflection at the soil-waveguide interface.

Secondly, since the waveguide is of finite dimensions, the frequency components in the seismic wave for which the wavelength is close to the bar diameter are likely to be diffracted causing additional attenuation. It is also important to note that the waveguide will be "transparent" for low frequencies (wavelengths greater than four times the bar diameter), and consequently only high frequencies will be coupled into the waveguide. Considering these factors, only a small part of the seismic signal will "cross" the soil-waveguide boundary. A simple calculation shows that, for typical densities and the velocity values mentioned earlier, at most only approximately 10% of the energy which reaches such a boundary will be coupled into the waveguide.

Thirdly, energy losses occur as the seismic signal propagates along the waveguide. Since in normal use the waveguide is coupled to the soil over its total length, each point of contact can be considered as a new "emitter" (according to Huygens' principle), and consequently the signal will be attenuated further. This effect could be minimized if the waveguide were not coupled over its full length (e.g. a certain portion was located within a suitable "isolation jacket") but in most cases this would be counterproductive since one of the alleged advantages of using waveguides is to increase the area of influence (Lord et al., 1982). Furthermore, since the attenuation through the waveguides itself is relatively low, reflections will occur at the ends of the waveguide further complicating data interpretation. If non-resonant type transducers are utilized (e.g., accelerators) the apparent transducer characteristics (composite characteristics of the transducer and the attached waveguide) will be very different from those of the transducer alone. Based on these factors, any frequency analysis of the recorded AE/MS signals would be prejudiced and resulting spectra could only be used for relative comparison of signals detected at the same time, using the same waveguide-transducer system.

Finally, since only one transducer is used with conventional waveguides, signals generated by relative motion between the waveguide and the surrounding media, and external seismic signals coupled into the waveguide, cannot be distinguished as to their point of origin along the waveguides. This problem prevents the data collected from single-transducer waveguides from being used for the purpose of AE/MS source location. However, this problem may be overcome using transducers located at both ends of the waveguide.

6.6.4   Geotechnical waveguide development

General concepts: In geotechnical projects (e.g. tunnel construction, slope modification and stabilization, mine roof support, etc.) it is important to have the capability of evaluating the current state of mechanical stability. AE/MS techniques are uniquely suitable to provide such real-time stability data. Conventional AE/MS monitoring normally employs a series of single element transducers located at various locations in or on the structure under study. For example, Figure 6.29a shows a typical competent rock mass, and an associated array of AE/MS monitoring transducers. In contrast Figure 29b illustrates the loss of direct propagation paths for two transducers (B and C) as a result of partial deterioration of the rock mass. In such a case the installed transducer array may become unsuitable for subsequent monitoring of the rock mass.

It is the author's opinion that the effect of rock mass deterioration on AE/MS monitoring capability may be greatly reduced by the use of mechanical waveguides. Figure 6.29c illustrates how this would be possible in the situation noted earlier in Figure 6.29a. Here a hole has been drilled into the rock mass traversing four rock layers, 2, 3, 4 and 5. As shown, a steel waveguide rod is installed in the hole, grouted in place, and a suitable transducer mounted on the outer end. Prior to rock mass deterioration, the acoustic energy associated with AE/MS sources, such as S1 and S2, are coupled into the steel waveguide and are detected by the attached transducer (A). When deterioration occurs, Figure 6.29d, the original direct path first arrivals are blocked by the development of

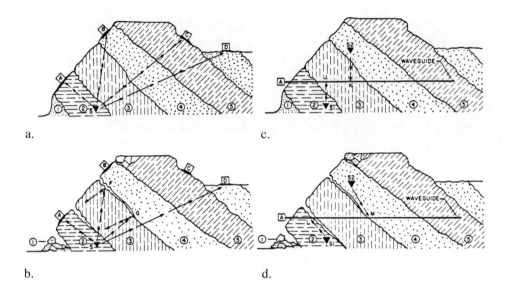

a.

c.

b.

d.

Figure 6.29. Effect of rock mass deterioration on the ability to detect AE/MS activity using surface transducers and waveguides (after Hardy et al., 1988). a. Rock mass prior to deterioration showing direct propagation paths between an AE/MS source (S) and a series of transducers (A, B, C and D). b. Rock mass after partial deterioration showing loss of direct propagation paths between an AE/MS source (S) and transducers B and C. c. Rock mass prior to deterioration showing direct propagation paths between two AE/MS source locations (S1 and S2) and a horizontal waveguide. d. Rock mass after partial deterioration showing other available propagation paths between two AE/MS source locations (S1 and S2) and a horizontal waveguide.

open fractures between blocks 2 and 3 and 3 and 4, nevertheless the same sources (S1 and S2) would still have propagation paths available to the waveguide and hence to the attached transducer. As a result, data from transducer A could still be analyzed to provide information on the general rate of AE/MS activity occurring in the rock mass.

A major problem with conventional waveguides is that they are unsuitable for AE/MS source location, since the actual point of impact of the AE/MS generated stress wave on the waveguide is not known. One possible means of improving the applicability of mechanical waveguides would be to mount a transducer on both ends, as shown in Figure 6.30. Here such a configuration will be known as a "dual-transducer waveguide". In this case AE/MS data from both transducers (A and B) are recorded. Analysis of this data using linear source location methods makes it possible to determine the AE/MS rate and also to define those regions of the slope where major AE/MS activity is occurring.

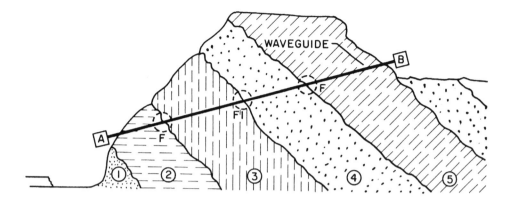

Figure 6.30. Dual-transducer waveguide configuration applied to rock mass monitoring (after Hardy, 1992).

It should be noted that both single- and dual-transducer waveguides are sensitive to AE/MS activity occurring within the rock mass. In particular, such waveguides are highly sensitive to mechanical instabilities occurring at boundaries between structural elements. For example in Figure 6.30, when rock mass deterioration begins, relative motion occurs between adjacent blocks, which is turn acoustically stimulates the mechanical waveguide at those interface areas designated F. As in the case of the single-transducer waveguides, even though serious deterioration of the mass occurs, AE/MS activity occurring in those regions through which the waveguide passes will be monitored.

Acoustic antennae and waveguides: According to a recent paper (Hardy, 1992), studies have been underway to investigate the use of mechanical waveguides and associated rock bolts (acoustic antennae) to monitor mine and tunnel roof stability. It should be noted, however, that the concepts developed in this application will be applicable to a variety of other geotechnical applications. Figure 6.31a illustrates a simplified form of the concept. Here a horizontal waveguide is attached to a series of in-place rock bolts, and suitable high frequency AE-transducers are located at the two ends of the waveguide. AE signals generated in the roof rock, due to mechanical instabilities in the rock and/or mechanical action on the roof bolts themselves, are transmitted down the roof bolts and injected into the horizontal waveguide. The use of an appropriate linear (1D) source location procedure to process the data from the two transducers makes it possible to determine where (i.e., at which rock bolt) the AE/MS signal was injected into the waveguide. As noted in Figure 6.31b, for a grouted bolt acoustic energy could be lost back to the rock mass in the attenuation zone and, due to coupling losses, where the acoustic antenna (rock bolt) is joined to the horizontal waveguide. For a mechanical bolt, problems

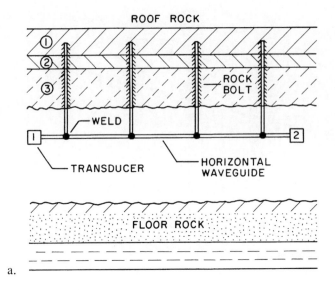

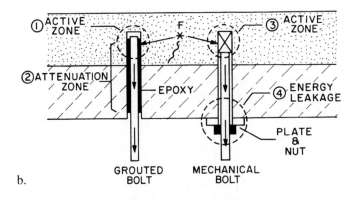

Figure 6.31. Concepts associated with AE/MS mine and tunnel roof monitoring system (after Hardy, 1992). a. Simplified form of mine roof monitoring system. b. Acoustic characteristics of grouted and mechanical bolts.

can also occur as a result of the energy leakage between the metal plate and nut (used to provide bolt reaction) and the associated rock.

In order to study the concept in more detail a concrete model (2′ × 2′ × 12′ long) containing a number of holes for mounting mechanical – and grouted – type rock bolts was constructed. To date, the acoustic behavior of a number of components of the system have been under study. These include the attenuation effects associated with the coupling of the rock bolts to the horizontal waveguide, the effect of torque on the behavior of mechanical bolts, the effect of grouting-length and bolt material on the behavior of grouted bolts, waveguide connection, and the importance of energy leakage. Detailed results on the above are available (Hardy, 1992) and studies are continuing.

Acoustic emission monitoring rod: Since the late 1980s Japanese researchers have been developing and utilizing a unique type of waveguide (Nakajima, 1988; Nakajima et al., 1995). In contrast to conventional passive waveguides, the so-called "acoustic emission monitoring rod" (AEMR) is an active waveguide in that distortion (bending and shear) of the waveguide generates internal AE/MS activity which is detected by AE-transducers attached by magnets to both the

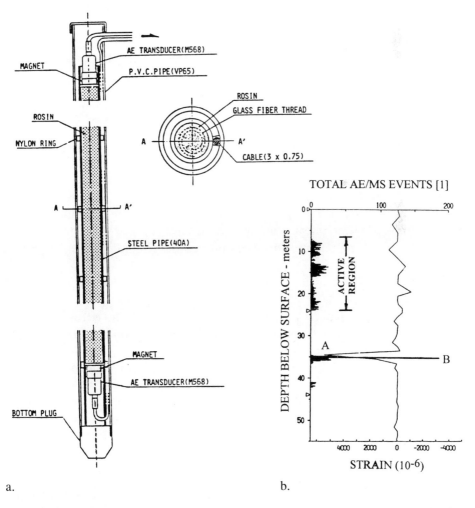

a.   b.

Figure 6.32. Acoustic emission monitoring rod and typical field results (after Nakajima et al., 1995). a. Structure of acoustic emission monitoring rod. b. Typical field results over 58-day period. ([1] total events per 0.2 m of rod length.)

top and bottom of the waveguide. Figure 6.32a illustrates the general structure of the unit.

The AEMR consists of an inner steel pipe (4.86 cm in diameter) containing a mixture of rosin and glass fibre with AE/MS transducers attached to both ends. The inner pipe is contained within an outer pvc pipe (7.60 cm in diameter) and centered in place by a series of nylon rings. The length of the overall unit appears to the approximately 2.5 m. Strains in the outer pipe were monitored by resistance-type strain gages cemented to the pipe at 0.5 m intervals.

The AEMR units have been applied successfully in landslide monitoring studies over a period of time greater than one year. Figure 6.32b shows typical strain and AE/MS data collected at one field site over a 58-day period. There is a good correlation between bending strain and AE/MS activity at a depth of approximately 35 m (strain A, AE/MS activity B), and over an extended active region ranging from approximately 8 m to 24 m. However, it is clear that AE/MS activity is a much more sensitive indicator of landslide instability than strain. The AEMR system appears to provide a very useful tool for monitoring landslides and stability of rock and soil slopes.

6.7   AMPLITUDE DISTRIBUTION ANALYSIS

The amplitude of AE/MS events observed for geologic materials under load have been found to vary over several orders of magnitude. In seismology the relationship between the cumulative number of events of magnitude M or greater in a given time interval n(M), and their magnitude M, has been found to obey the following relationships (Guttenberg and Richter, 1949):

$$\log n\,(M) = a + b\,(8 - M) \qquad\qquad (Eq.6.4)$$

It should be noted that in seismology the term magnitude is related to the logarithm of the observed amplitude of the recorded seismic event. Some 30 years ago the above relationship was found to apply to small earthquakes as well as to AE/MS data observed in stressed specimens of geologic materials (e.g. Knill et al., 1968; Scholz, 1968; Mae and Nakao, 1968; Suzuki, 1953, 1954; Mogi, 1962). Rather than applying Equation 6.4 directly, Scholtz (1968) presented his experimental data in terms of the Ishimoto-Iida statistical relationship (Ishimoto and Iida, 1939), namely:

$$N(A)dA = KA^{-m}dA \qquad\qquad (Eq.\ 6.5)$$

where N(A) is the amplitude frequency, namely the number of events having amplitudes in the range A to A + dA, and K and m are constants. Scholz also noted that there is a relationship between the constant b and m in Equations 6.4 and 6.5, namely:

$$b = m - 1 \qquad\qquad (Eq.\ 6.6)$$

Figure 6.33 illustrates the variation of N(A) with amplitude for tests on Westerly Granite at two different levels of uniaxial compressive stress. Two important features are indicated in this figure, namely the value of b varies with stress level, and the data deviate significantly from a straight line for large amplitude events. Scholz (1968) observed that the factor b (Eq. 6.4) varied inversely with the level of applied stress for a number of rock types investigated under uniaxial compression, as illustrated in Figure 6.34. Such behavior indicates that with increasing stress level a higher percentage of the observed events have greater amplitude.

In contrast, however, studies by Sun et al. (1991) on Berea Sandstone, stressed at a uniform rate to failure show a different trend. Here computed b-values ranged from 1.04 to 1.78, however, there was no consistant relationship between b-values and applied stress.

Amplitude distribution analysis has been used extensively in AE/MS studies of a variety of materials including metals, ceramics, composites and geological

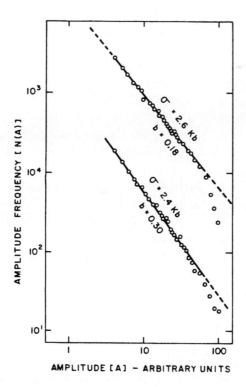

Figure 6.33. Amplitude frequency versus amplitude for AE/MS events observed during two experiments on westerly granite under uniaxial compression (after Scholz, 1968). (1 kb = 100 MPa.)

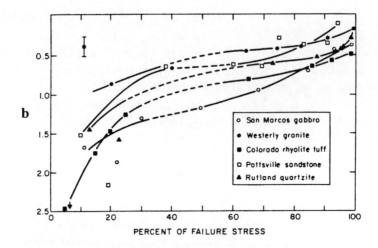

Figure 6.34. Variation of b-values with uniaxial compressive stress for several rock types (after Scholz, 1968).

materials. Gibowicz and Kijko (1994) provide a review of the application of amplitude distribution analysis to a variety of earthquake and rockburst studies. Trombik and Zuberek (1977) have used the technique for the study of destressing blasting in deep Polish coal mines. Later studies by Zuberek (1989) consider the application of the Gumbel asymyptotic extreme value distributions in the study of AE/MS amplitude distribution data.

Dürr and Meister (1984) discuss the use of AE/MS techniques in the study of pillar deformation in a German salt mine (Asse) and the application of amplitude distribution analysis. A later paper (Meister et al., 1998) expands the length of the Asse Mine study to a 13-year period. During this period, based on data from two transducers, there was a general increase in the observed AE/MS rate from zero to an average value of 4000 events per hour. Yearly cyclic variations in AE/MS event rate were also noted which appear closely related to the mine relative humidity level. The results of amplitude distribution analysis of a large number of events for a one day period each year exhibit a linear trend, with regression coefficients varying from 0.913 to 0.990, b-values in the range $0.065 \leq b \leq 0.098$, and no consistent, well defined shift in b-values with time. This suggests that there has been no significant increase in structural loading, as earlier studies (Scholz, 1968) suggest that b-values should decrease with increasing stress. However, as in the case for Berea Sandstone noted earlier, salt may show a different trend.

Holub (1998) recently published detailed data on b-value changes during coal mining in the Ostrava-Kurvina coal basin, Czech Republic. His findings confirmed that b-values are inversely proportional to the level of induced seismicity generated by various mining activities. For example, activities such as development work, reduced coal extraction, and interruptions and/or stoppage in mining, resulted in a decrease in seismic activity and higher b-values. In contrast fully developed longwall panels and/or developing panels exposed to extreme stress conditions resulted in increased rock mass loading, which was reflected in lower b-values.

6.8 PATTERN RECOGNITION

The use of pattern recognition and related neural networks have become an increasingly important component of AE/MS data analysis. In the geotechnical area Shen (1996, 1998) has recently completed a detailed study of cutting and breakage of coal and made extensive use of pattern recognition in the analysis of experimental data. The following section will include a brief outline of pattern recognition concepts (Shen, 1996). Those who are interested in further details should refer to Bow (1992), Fu (1982), and Chen (1973).

6.8.1 Introduction

Pattern recognition usually involves the classification and/or description of a set of processes or events. The process or events with some similar properties are grouped into a class. The total number of pattern classes in a particular problem is often determined by the particular application in mind. For example, consider the problem of English character recognition; here we have a problem with 26 classes.

The many different mathematical techniques used to solve pattern recognition problems may be separated into two general groups, namely, the syntactic approach and the statistical approach (Fu, 1982). In the syntactic approach, each pattern is expressed as a composition of its components, called sub-patterns and pattern primitives. Some typical applications of the syntactic pattern recognition are image recognition, character recognition, fingerprint recognition, and speech recognition. In the statistical approach, a set of characteristics measurements, called features, are extracted from the patterns; the recognition of each pattern assignment to a pattern class is usually made by partitioning the feature space.

Over the past three decades, statistical methods have played an important role in the development of pattern recognition techniques. The statistical decision theory and related fields have been a forum where significant theoretical advances and innovations have taken place. These, in turn, exerted a strong impact on pattern recognition applications including seismic signal recognition, ultrasonic nondestructive evaluation, AE/MS signal recognition, underwater acoustic waveform recognition, remote sensing, etc.

6.8.2   Statistical pattern recognition methodology

In general, AE/MS signals are randomly occurring transients whose characteristics depend on the mechanical properties of the laboratory specimen or field structure under study, and on the degree and type of instability involved. Normally raw analog AE/MS data is received in the time domain, and this continuous waveform must first be digitized to provide a series of data points which represent the first step in the statistical pattern recognition process. Each data point may provide a characteristic feature of the captured signal in the time domain. such a characteristic feature, in turn, carries information about the microstructure of the material within which the associated stress wave has been propagating. For the purpose of pattern recognition analysis, it is necessary to form a Euclidean space, where each of the coordinate axes are represented by the characteristic features obtained from the AE/MS signal. The large number of characteristic features from such a signal would constitute a multi-dimensional Euclidean space. Such high dimensionality makes pattern recognition problems rather difficult. For simplification, therefore; it is necessary to extract, from the primitive measurements, only the most important characteristic features. Each of these features carry a small, but significant, amount of information for classification purposes and are usually selected according to the "physics of the problem".

Feature selection is achieved through a process known as "feature extraction" which is the second step in the pattern recognition process. The pertinent feature extraction process would involve "mapping" from the primitive "n-dimensional" space to a lower dimensional space, without losing any of the significant information characteristic of the pattern class. This is often referred to as preprocessing. For example, typical features for AE/MS analysis might be the largest peak amplitude and its position. If a complete set of discriminatory features for each pattern class can be determined from the measured data, then the recognition and classification of patterns will present little difficulty and automatic recognition may be reduced to a simple matching process. For this purpose algorithms have been developed for the extraction of representative features from primitive measurements.

The third step in the pattern recognition process involves an "optimum decision" procedure that is needed in the classification process. After the observed data from the patterns to be recognized have been expressed in the form of pattern points or measurement vectors in pattern space, one must then decide to which pattern class these data may belong. On assuming, for instance, that a "machine" is to be designed to recognize M different pattern classes denoted by w_1, w_2, ..., w_M, then the pattern space can be considered as consisting of M regions, each of which encloses the pattern points of a class. The recognition problem can then be viewed as that of generating the decision boundaries which separate the M pattern classes on the basis of the observed measurement vectors.

Since the principal function of a pattern recognition system is to yield decisions concerning the class membership of patterns, it is necessary to establish some rules upon which to base such decisions. One important approach to this problem is the use of "decision functions". As an example of a decision function, Figure 6.35a illustrates two pattern classes, separated by a line in a two dimensional feature space. In an n-dimensional case, a general linear decision function may be expressed in the form

$$d(x) = w_1 x_1 + w_2 x_2 + \ldots w_n x_n + w_{n+1} \qquad \text{(Eq. 6.7)}$$

where d is a distance measure definable over the feature space, x_1, x_2, ..., x_n are components associated with the pattern vector x in an n-dimensional feature space, and w_1, w_2, ..., w_n, w_{n+1} are constant parameters. A classifier using such a linear function is referred as a "linear discriminant" classifier.

Another way of establishing a measure of similarity between pattern vectors is by determining their proximity to their respective class. This is illustrated in Figure 6.35b which shows that a vector (pattern point) x belongs to a class w_2 on the basis that it is closer to the pattern of this class. This method of pattern classification, based on the proximity concept or distance function, can be expected to yield practical and satisfactory results only when the pattern classes tend to have clustering properties, i.e. the intra-class pattern vectors group together and are well separated from those belonging to the other classes as illustrated in Figure 6.35b. Other commonly used classifiers include "nearest neighbor" and "empirical Bayesian".

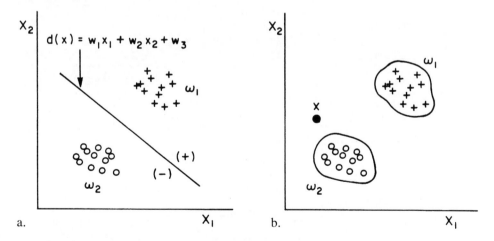

Figure 6.35. Examples of two-feature signal classifiers (after Shen, 1996). (ω_1 and ω_2 denote pattern classes 1 and 2. X_1 and X_2 denote rank No. 1 and No. 2 features.) a. Linear discriminant classifier. b. Proximity concept classifier.

The pattern classification process, considered in this text, depends on a priori knowledge of the pattern class boundaries. Thus, two processes of "training" and "evaluation" are carried out. The input data are first split into two separate files in the training procedure. This is done after normalizing the data with respect to their variance. The first file that contains the values of the normalized feature vectors is used to create the boundary between the pattern classes. Once the boundary is established, a classifier is designed and then evaluated using the data stored in the second file. The results of the training and evaluation processes are expressed as a percentage of the success in forming distinct clusters in the feature space from the known pattern vectors. The efficiency of a particular classifier can then be tested using a raw data set taken from an unknown sample creating a pattern vector in the feature space that needs to be classified to its respective class. A flow chart illustrating the training and evaluation procedure associated with a typical pattern recognition system is shown in Figure 6.36.

A variety of pattern recognition programs are presently available. A number of workers, including the author, have had excellent experience using the commercially available program known as ICEPAK (*I*ntelligent *C*lassifier *E*ngineering *P*ackage). ICEPAK is a computer software package specifically designed for recognition of waveforms such as AE/MS signals (Hay and Chan, 1994, 1995). Data files previously generated by various acquisition systems may contain either a single event or multiple events in a variety of data formats. The

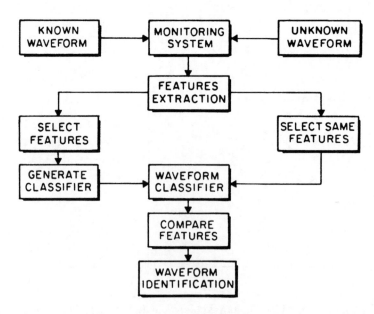

Figure 6.36. Flow chart showing the training and evaluation procedure associated with a typical pattern recognition system (after Shen, 1996).

system supports signals, up to 4096 data points in length per record, however, a series of long AE/MS signals can be represented by a group of records. The classifier uses various properties or features to characterize the waveform. A feature extraction process automatically extracts 108 predefined features from each record, and these feature values are stored in a feature file. Before a classifier can be designed or developed, the feature files must be normalized. The normalization process maps the actual feature values or measurements, such as peak amplitudes, areas, slope values, etc., into dimensionless numbers ranging from –1 to 1. ICEPAK provides five pattern classifiers in the system, namely, linear discriminant function classifier (LD), K-nearest neighbor classifier (KNN), Empirical Bayesian classifier (EB), minimum distance classifier (MM), and neural network classifier (NN).

6.8.3 Acoustic emission/microseismic applications

Pattern recognition techniques have been applied to AE/MS data processing since the early 1980s. Although with the rapid development of more and more powerful personal computers, and "user friendly" pattern recognition software, general AE/MS applications have been increasing rapidly. The papers by Murthy et al. (1987), Ohtsu and Ono (1987), and Ono and Huang (1994) provide an excellent review of the literature over the last 15 years and include more than 50 related references.

Use of pattern recognition techniques in AE/MS studies of geologic structures and materials have been relatively limited. Literature of interest here include papers by Hay and Chan (1995), Feknous et al., (1995) and Shen (1996, 1998). A brief review of Shen's studies on cutting and breakage of coal, and the analysis of the associated AE/MS activity using pattern recognition, is presented in Shen (1996, 1998).

6.8.4 Neural networks

As noted by Shen (1996), a frustrating characteristic of conventional computers is the literal, precise inputs required to produce the desired output. Neural networks, on the other hand, can accommodate variations in their input and still produce correct output. Broadly stated, neural networks map vectors from one space into another. Vectors, in their generalized mathematical sense, can be composed of elements with a wide range of characteristics.

Rather than being programmed, neural networks are trained by example. Just as children need know nothing about comparative physiology to recognize their family members, programmers need not provide neural networks with quantitative descriptions of objects being recognized, nor sets of logical criteria to distinguish such objects from similar objects. Instead, examples of objects (faces or scenes) along with their identifications (mother or lake) are shown to the neural network. It memorizes these by altering values in its weight matrix, and produces the proper response when an object is seen again. The real world rarely presents information with the precision required by a computer program, and neural networks accomplish the needed generalization by virtue of their structure rather than through elaborate programming. In recent years neural networks have been utilized at an increasing rate to process AE/MS data. papers by Grabec (1992), Grabec et al. (1991), Ida et al. (1991), and Yuki and Homma (1992) illustrate a variety of applications.

Those readers who are interested in further details on the concept of neural networks should refer to other sources such as Dayhof (1990), Hay and Chen (1994), Simpson (1989), Studt (1991), Wasserman (1990), and Wasserman and Schwartz (1988).

6.8.5   Wavelet transform

A recent development of particular interest in the pattern recognition area is the use of the so-called "wavelet transform" for analysis of AE/MS signals. According to Suzuki et al. (1996), the wavelet transform (WT) allows the determination of frequency spectrum as a function of time using short waveform segments or wavelets. The authors indicate that the procedure provides more informative characterization of AE/MS signals than the power-density spectra obtained from the conventional Fourier transform. Suzuki et al. (1996) provide a review of the WT concept, include a computer program for performing WT analysis, using the Gabor Wavelet, and illustrate its application to the analysis of AE/MS data from mechanical tests on a glass-fibre reinforced composite (GFRP). Although limited at present in its application in geotechnical studies the wavelet transform appears to be a useful tool for analysis of AE/MS data from such studies.

6.9   SOURCE PARAMETER ANALYSIS

A number of researchers, in particular those with a geophysics (seismology) background, have made use of source parameter analysis to classify the wide range of observed AE/MS events. Based on Brune (1970) three parameters are often considered, namely: the radius of the rupture surface, the seismic moment and the stress drop. These parameters are defined as follows,

$$r = 1.97 \, V_p / 2\pi f_c \qquad\qquad \text{(Eq. 6.8)}$$

$$M_o = 4\pi R \rho V_p^3 \, \Omega_o \qquad\qquad \text{(Eq. 6.9)}$$

and

$$\Delta\sigma = 7M_o / 16r^3 \qquad\qquad \text{(Eq. 6.10)}$$

where r is the source radius, V_p is the p-wave velocity, f_c is the "corner frequency" for compressional waves, M_o is the seismic moment, R is the transducer (sensor) to hypercentre distance, r is the rock density, Ω_o is the low frequency spectral level and $\Delta\sigma$ is the stress drop. Similar relationships hold for shear waves. Further details relative to the basis of the preceding equations are given by Brune (1970, 1971) and Greenfield (1977). Figure 6.37 illustrates S-wave log displacement spectral density versus log frequency for an idealized AE/MS event, and the procedure for evaluation of associated Ω_o and f_c parameters.

The application of source parameter analysis in geotechnical studies is considered in a number of papers including those on hydraulic fracturing (Baria et al., 1989; Power, 1977; Talebi et al., 1989), subsidence due to solution mining (Wong et al., 1989), stability of salt caverns used for oil storage (Albright and Pearson, 1984), and nuclear waste storage site stability (Majer et al. 1984). For example, Figure 6.38 shows the log $\Omega(w)$ versus frequency data for two AE/MS events associated with depressurization studies at an oil storage cavern located in rock salt (Albright and Pearson, 1984). Table 6.2 presents typical source parameter values computed for a number of events monitored during this study.

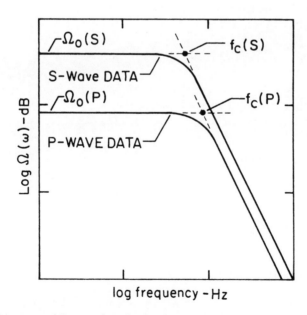

Figure 6.37. P-wave and S-wave log displacement spectral density versus log frequency for an idealized AE/MS event (after Greenfield, 1977). (Associated values for Ω_o and f_c are indicated.)

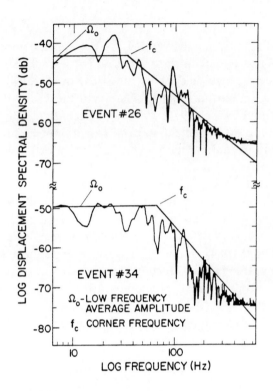

Figure 6.38. S-wave log displacement spectral density versus log frequency data and associated values Ω_o and f_c for two events monitored at a salt cavern oil storage facility (after Albright and Pearson, 1984).

Table 6.2. Typical source parameter values for AE/MS events monitored at a salt cavern oil storage facility (after Albright and Pearson, 1984).

Event Number	$f_c(1)$	r	M_o	$\Delta\sigma$
	Hz	m	10^{+7} Joule	M^{-3} MPa
32	52	19	3	2
26	33	29	15	3
8	61	16	3	4
34	66	14	6	8
40	66	15	15	22
24	39	25	77	22

(1) f_c - corner frequency, r - source radius, M_o - seismic moment, and $\Delta\sigma$ - stress drop.

6.10 MOMENT TENSOR ANALYSIS

Detailed discussion of moment tensor analysis (MTA) is beyond the scope of the current text. However, a brief outline of the topic will be included here, along with those references necessary for the reader to become more familiar with the topic. A recent paper by Manthei et al. (1998) provides an excellent outline of the historical background of the topic and its application to the subject of induced cracking in rock salt.

Basically, MTA involves the waveform analysis of signals generated by thermal and mechanical instabilities. According to Gibowicz and Kijko (1994), if a general body force representing a seismic source can be expressed as a linear combination of couples with moments M_{ij}, then the displacements caused by this force can be written as

$$U_k = M_{ij} * G_{ki,j} \qquad \text{(Eq. 6.11)}$$

where M_{ij} is a set of nine terms and is known as the "moment tensor" of the source, $G_{ki,j}$ is the spatial derivative of the Green's function, and $*$ denotes convolution. Studies have shown that fracture orientation and type (tensile/shear) can be determined from moment tensor analysis.

In seismology, moment tensors for larger earthquakes have been determined using waveform inversion based on the Centroid Moment Tensor method (CMT) of Dziewonski et al. (1981). This method determines simultaneously the six independent components of the moment tensor and the coordinates of the centroid. Ebel and Bonjer (1990) performed moment tensor inversion of small earthquakes in southwestern Germany. Gibowicz (1993) has used the moment tensor method to study the mechanism of events caused by mining. Ohtsu (1991) applied MTA to the analysis of data from a hydrofracturing study, and more recently Ohtsu (1995) developed a computer code named SiGMA (simplified Green's function for moment tensor analysis) which utilizes the P-wave amplitudes for characterization of AE/MS events. In other studies Ohtsu and others have applied moment tensor analysis to the study of cracking in reinforced concrete (Ohtsu et al., 1994; Yuhama et al., 1994).

6.11. KAISER EFFECT

6.11.1 Introduction

Dr. Joseph Kaiser (1953), a German researcher, was the first person to observe the phenomenon which now bears his name. The discovery was made during a study of the acoustic emission (AE) response of metals, which Kaiser conducted in the early 1950s, and indicated that materials retain a "memory" of previously applied stresses. Over the last five decades, continued research on the Kaiser effect in metals, ceramics and composites has enabled the development of a number of practical applications. In the geotechnical area (Montoto and Hardy, 1991), the major research interest in recent years has been associated with the use of the Kaiser effect as a means of evaluating in-situ stress. As a result the majority of the published material is concerned with the effects of mechanical loading, although a number of recent papers have been concerned with thermal loading.

Papers by Hardy et al. (1989) and Momayez et al. (1990) provide a general review of the Kaiser effect research carried out on geologic materials through the late 1980s. Furthermore a considerable number of papers dealing with specific topics appear in the literature. These include studies on a variety of rock types and soil. The literature also indicates that the Kaiser effect occurs in concrete, a material similar in many aspects to rocks, under thermal (Diederichs et al., 1983) as well as mechanical (Mlaker et al., 1984) loading. Interest in the Kaiser effect in geologic and related materials, like concrete, has increased rapidly in recent years.

Research programs of a continuing nature are underway in a number of countries. Japan, where studies in Electric Power Industry (CRIEPI) reactivated the current worldwide interest in the Kaiser effect (Kanagawa et al., 1981), is the most active. Detailed studies are also underway by the OYO Corporation, and the Ohbayashi Corporation in cooperation with a number of Japanese Universities (Michihiro et al., 1989). These include Tohoku University (Kojima and Matsuki, 1990) and Tokyo University (Yoshikawa and Mogi, 1989). A recent paper by Seto et al. (1997) deals with Kaiser effect studies of coal.

In recent years, detailed Kaiser effect studies have been underway in the USA and Canada. In the USA these include studies at the U.S. Bureau of Mines (Fridel and Thill, 1990), Penn State University (Hardy, 1998b; Hardy et al., 1989; Shen, 1995), the Sandia National Laboratories (Holcomb and Martin, 1985), and the Mackay School of Mines (Watters and Roberts, 1998). During the early 1980s considerable research was underway at Drexel University, both on rock and soil (McElroy, 1985), however, it appears these studies are presently inactive. In Canada, studies have been underway at McGill University (Momayez and Hassani, 1998; Momayez et al., 1990), Sherbrooke University (Fekmous et al., 1995) and Toronto University (Hughson and Crawford, 1986).

Kaiser effect studies under mechanical loading have also been underway in Russia (Lavrov, 1997; Chkowratnik and Lavrov, 1997), India (Mukherjee et al., 1998), the United Kingdom (Stuart et al., 1995; Barr, 1993), Sweden (Li, 1998), Australia (Wood et al., 1995; Wang, 1999), and China (Zhang et al., 1998). Finally, a limited number studies under both mechanical and thermal loading have been reported in the United Kingdom (Atkinson et al., 1984), Spain (Montoto and Hardy, 1991; Montoto et al., 1989; Montoto et al., 1998) and Poland (Zuberek et al., 1998).

6.11.2 Measurement and analysis

The Kaiser effect due to mechanical loading is illustrated in Figure 6.39, where a rock specimen is subjected to two cycles of loading. As illustrated in Figure 6.39b, in the first load cycle, stress is applied to the specimen at a constant rate up to a value of σ_{max} and then reduced to zero. In the second cycle, stress is increased in a similar fashion, however, the previous maximum stress (σ_{max}) is exceeded. During each cycle, AE/MS activity is monitored and accumulated AE/MS counts (or events) recorded as a function of applied stress. Figure 6.39b shows the cumulative number of AE/MS counts generated within the specimen during each of the two load cycles. It can be seen that AE/MS activity is generated at all stress levels during the first cycle. However, during the second load cycle little or no activity is generated until the maximum stress (σ_{max}) attained in the first load cycle is exceeded. Thus, the Kaiser effect can be defined as the absence of AE activity at stress levels below a previously applied maximum stress.

In geotechnical applications, rock cores would first be obtained from a field structure where in-situ stress data was required. These cores would then be prepared into test specimens and loaded in the laboratory over a suitable stress range, while the associated AE/MS data was recorded. Analysis of the resulting

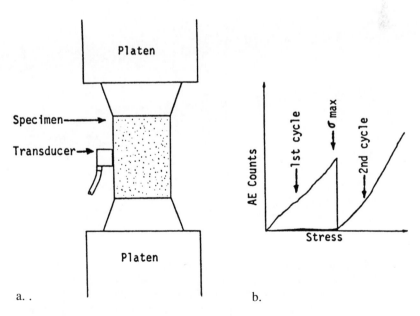

Figure 6.39. A simple laboratory arrangement and typical data associated with Kaiser effect tests (after Hardy et al., 1989). a. Testing arrangement. b. Typical test data.

AE/MS data would provide the Kaiser stress (σ_k) values, which are assumed to represent the in-situ stress at those points in the field structure where the rock cores were obtained. Figure 6.40a shows data for a uniaxial test on Indiana Limestone, originally loaded to a stress of 4054 psi (Hardy et al., 1989). Here incremental counts, namely the number of counts for each 500 psi (3.43 mPa) increase in the reload stress, are shown plotted against the applied stress level. Linear regression techniques were then applied to the data and two lines (L1 and L2) fitted to the data below and above the "apparent" Kaiser stress level. The intersection of lines L1 and L2 was then projected to the stress axis to define the actual Kaiser stress (σ_k). In general, based on controlled experiments, the computed Kaiser stress may not be exactly equal to the previous maximum applied stress. The deviation is normally indicated by the value of the so-called "Felicity ratio" (Anon., 1992). This is defined as the ratio of computed Kaiser stress (σ_k) and the previous maximum prestress (σ_p), namely: FR = σ_k/σ_p. For the data in Figure 6.40a, FR = 1.02, and in Figure 6.40b, FR = BC/BA = 0.26.

AE/MS studies under thermal loading have also been carried out, and suggest that the existence of a "thermal Kaiser effect" (see Montoto and Hardy, 1991). Figure 6.41 illustrates typical data from thermal cycling tests on Westerly Granite (Atkinson et al., 1984).

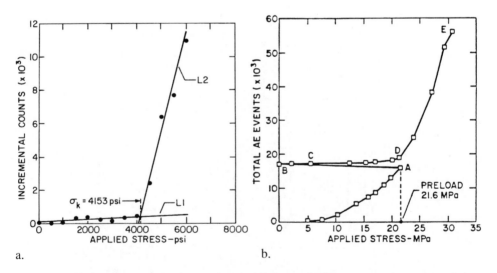

a.   b.

Figure 6.40. Experimental data illustrating the Kaiser effect and the felicity ratio for specimens of limestone and potash. a. Incremental counts versus stress for Indiana limestone specimen 3 (after Hardy et al., 1989). (uniaxial preload σ_p = 4054 psi, Kaiser stress σ_k = 4153 psi. 1 MPa = 145.99 psi.) b. Total events versus stress for potash, test PCS-2 (after Mottahead and Vance, 1989). (uniaxial preload σ_p = 3153; Kaiser stress σ_k = 819 psi (point c). 1 MPa = 145.99 psi.)

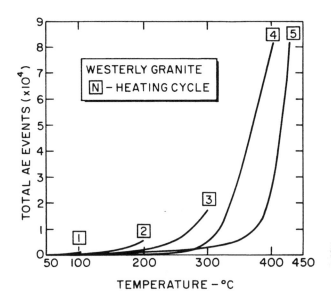

Figure 6.41. Typical AE/MS data from a thermal cycling test on western granite illustrating the thermal Kaiser effect (after Atkinson et al., 1984).

6.11.3 Mechanisms

In "hard rocks" AE/MS activity is assumed to be the result of such mechanisms as crack initiation and propagation, plastic deformation of individual grains, and frictional sliding due to intergranular reorientation prior to and following the initiation of dilatancy. Generally it is assumed that plastic deformation and intergranular reorientation at lower stresses are minimal in comparison to crack initiation and propagation. At stress levels where dilatancy occurs, fracture and intergranular reorientation become the major AE/MS mechanisms. Furthermore, once dilatancy occurs it would appear that limited "healing" of the material will occur even under reduced stresses. As a result of this it is assumed that for hard rocks the Kaiser effect breaks down, and the Felicity ratio decreases, when the applied stress rises above the point at which dilatancy occurs.

Li (1998) states that AE/MS activity emitted from compressively loaded rock is mainly caused by fracture within the rock. He notes that it is well known that (brittle) rocks contain a large number of pre-existing rocks. During compressive loading of such rocks, two mechanisms are active, namely: frictional movement between pre-existing crack surfaces, and fracture of intact material. Using the above, Li develops fracture models for use in one-, two-, and three-dimensional stress space (Li, 1995; Li and Nordlund, 1993a,b). These models were examined using data from uniaxial, biaxial and triaxial laboratory tests.

In contrast, in "soft rocks" such as salt and potash, the major AE/MS and deformation mechanism is clearly plastic deformation, although the other mechanisms noted earlier are certainly active. Recent studies (Hardy, 1998) have clearly indicated that polycrystalline salt exhibits a reliable Kaiser effect at stress levels well above that where dilatancy first occurs.

6.11.4 Current status and required studies

A general review of the Kaiser effect literature indicates the following areas of concern, namely:
1. *Specimen environment*: the effect of temperature and moisture conditions on test specimens during the pretest interval, namely the time between removal of the specimen from the in-situ stress field and subsequent testing.
2. *Memory loss*: the decay of the Kaiser effect with time during the pretest interval.
3. *Effects of test procedure*: the effect of such factors as specimen shape, end cap design, rate of loading and AE monitoring system gain.
4. *Determination of Kaiser stress level*: the need for an objective means for determining the Kaiser stress levels from the laboratory data.
5. *Multiaxial stress states*: the effect of biaxial and triaxial stress states on the values of the Kaiser stresses determined using uniaxial reloading tests.

Nevertheless, studies to date indicate that the Kaiser effect should provide the basis for the development of a practical procedure for the investigation of the stress fields induced in geologic materials and related materials such as concrete, as a result of both mechanical and thermal loading. Interest in the Kaiser effect has increased rapidly in recent years and considerable laboratory data is now becoming available. This data, however, exhibits a wide range of anomalies which must be investigated and resolved.

It is felt that considerable fundamental research is required to better understand the basic mechanisms responsible for the Kaiser effect. These include both microscopic and analytical studies. Until this is accomplished, progress in utilizing the Kaiser effect to solve practical geotechnical problems will be seriously delayed.

6.12 RELATED NDT TECHNIQUES

There are a wide range of non-destructive testing (NDT) techniques available for the laboratory and field investigation of geologic materials. These include well known techniques such as sonic/seismic transmission, refraction and tomography, which are active techniques and normally utilize artificial sources. In contrast the acoustic emission/microseismic technique is a passive technique utilizing sources generated by deformation and/or failure mechanisms within the material, when subjected to appropriate loading.

In recent years three additional NDT techniques, applicable to geologic materials, have developed, namely: acousto-ultrasonics, modal analysis and modal acoustic emission. To date these techniques have not been utilized extensively in the study of geologic materials. However, for completeness, a brief outline of each of the three techniques will be included here.

6.12.1 Acousto-ultrasonics

According to Drouillard and Vary (1994) acousto-ultrasonics (AU) was conceived as "a practical nondestructive technique for characterizing a selected volume of material relative to its properties and defect conditions." A recent review of the subject by these authors provides an excellent outline of the historical development of the subject, aand details on the associated instrumentation and typical applications. In essence AU is an active technique involving the application of a mechanical vibration to a test piece (specimen or structure). This vibration interacts with a selected volume of the test material and the modified vibration is monitored and processed to display various internal characteristics of the test piece.

Figure 6.42a illustrates a block diagram of an early AU system (ca. 1977). Here a sending transducer (transmitter), driven by a ultrasonic pulser and associated continuous wave oscillator, introduces mechanical vibrations into the test specimen. These input signals propagate through the specimen, undergoing modifications due to attenuation, dispersion, induced micro-level structural instabilities, etc. A pick-up cone and attached receiving transducer, spaced a specific distance away from the transmitter, detects the modified input signal which is processed by an associated monitoring system. As indicated in Figure 6.42a the monitoring facility shown appears to be a parametric AE/MS laboratory monitoring system of the type discussed earlier in Chapter 4. Here the presence or absence of activity, and in some cases signal amplitude, were the only evaluation parameters. During testing, various regions of the test specimen or structure could be investigated by selective positioning of the transmitted-receiver pair.

Figure 6.42b illustrates a block diagram of a typical present day AU monitoring facility; which incorporates a computer-based monitoring system. This allows full wave signal collection and sophisticated processing to generate a wide range of evaluation parameters.

To date, the AU technique has been applied with considerable success to the study of wood and fibre reinforced composites. However, it has not been applied extensively to the study of geologic materials or structures. Nevertheless tentative geotechnical applications are evident, namely: structural stability evaluation of mine and tunnel roofs, location of bed separation in rock

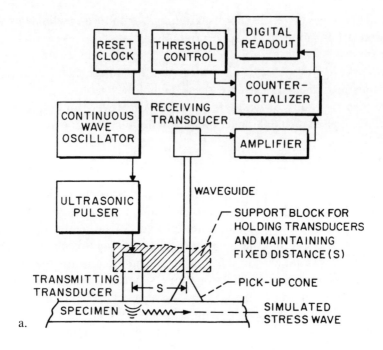

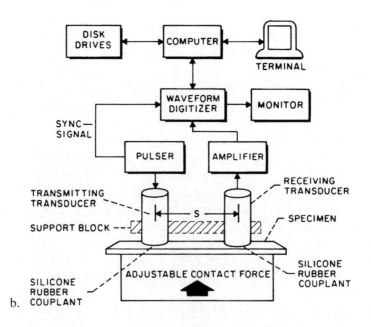

Figure 6.42. Block diagrams of early and current acoustic-ultrasonic (AU) monitoring facilities (after Drouillard and Vary, 1994). a. Early AU system with parametric monitoring system. b. Current AU system with computer-based monitoring system.

specimens and geotechnical structures, and detection of debonding in multi-layer concrete and asphalt structures such as roads and highway bridges. The paper by Drouillard and Vary (1994) includes a number of useful references on the AU technique. Additional information is available in publications by Anon. (1996d) , and Beall and Vary (1994).

6.12.2 Modal analysis

During the 1970s, a new branch of vibration engineering was developed, encompassing the most powerful aspects of both experimental and analytical techniques (Ewins, 1986). The resulting technology, which involves an active source, has become known as "experimental modal analysis" or "modal testing". It has shown great potential in assessing the characteristics of materials, and engineering structures by evaluation of their dynamic behavior.

A brief description of the modal analysis technique is presented here relative to studies on a large plate-like specimen (e.g. 0.6 m × 0.4 m × 0.028 m) as shown in Figure 6.43. The surface of the plate was first subdivided by a series of lines parallel to the x- and y-axis and intersections of these lines were numbered and are denoted as "test points". Prior to testing the plate was suspended from one edge to provide a free-boundary condition, and a suitable transducer (accelerometer) was attached to a selected test point.

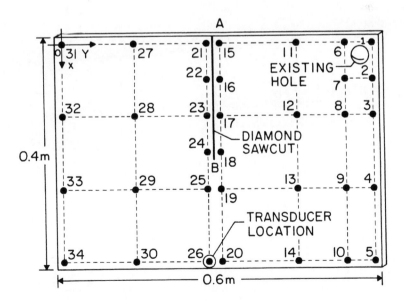

Figure 6.43. Laboratory specimen setup to conduct modal analysis tests (after Sun, 1991). (Solid dots are test points. Transducer is located at test point 26. Induced defect (saw cut) AB.)

A standard test procedure, the Broad-Band Impact Test, was utilized. This involved using a special instrumented hammer to impact each of the (33) test points, collecting the data received by the transducer (located at test point 26) and by the instrumented hammer, and processing this data. Analysis was accomplished using a dedicated computer-based dual-channel signal analyzer (WAVEPAC system) to provide the transfer function (TF) for each pair of input-output points, using a fast Fourier transform. A number of repeat impacts (≈ 30) were made at each point and the processed data was recorded on floppy disks. Figure 6.44 illustrates the typical instrumentation utilized.

The collected data may then be further processed to provide "mode shapes" and "modal parameters" (such as the damping coefficient), which describe the detailed vibration characteristics of the test specimen. If structural damage occurs in the specimen, data obtained from a subsequent test will yield different mode shapes and modal parameters, and these differences may be utilized to determine the location and severity of the structural damage.

Sun and Hardy (1990) describe modal analysis laboratory studies on a plate-like specimen of soapstone similar to that shown earlier in Figure 6.43. After test data was collected for the intact specimen, a cut was made through the plate (line AB) and modal analysis tests were repeated on the damaged plate. Figure 6.45 illustrates the various mode shapes observed before and after the cut. It is noted that the cut has clearly modified the mode shapes, inducing additional modes as well as eliminating certain modes observed in the undamaged plate. The modal parameters generally did not change for a particular mode shape, showing that although the structural behavior changed, the material properties did not change.

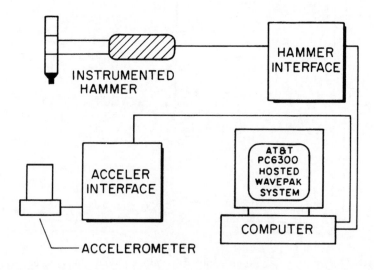

Figure 6.44. Block diagram of instrumentation used in laboratory modal analysis test.

The successful application of modal analysis techniques to laboratory tests on geologic material encouraged extension of the technique to a field scale study (Sun and Hardy, 1992). The test site selected was a potentially unstable rock slope located adjacent to a highway road cut, close to the pavement, and subjected to heavy traffic vibrations. The main object of the study was to investigate the behavior of a large rock plate, approximately 2.4 m × 2.1 m × 0.5 m thick, resting on a planar slope inclined at an angle of approximately 50°.

During a three-month period the block was modified several times to provide defect development. At each stage of modification, modal testing was conducted. First of all, a thorough statistical evaluation of the data collected during the modal testing was performed. Analysis proved the data to be random and transient, with a bell-shaped distribution. The system stability, time-invariance, and physical reliability were also demonstrated by analysis of the collected data. Slight nonlinearity was observed during the experiments and this was compensated for by limiting impact level. Subsequent data analysis indicated that the applicability of experimental modal analysis to the geotechnical structure was justified. Modal behavior was examined after each stage of block modification to show the influence of defect development. Variations of mode shapes were found both visually and analytically after each

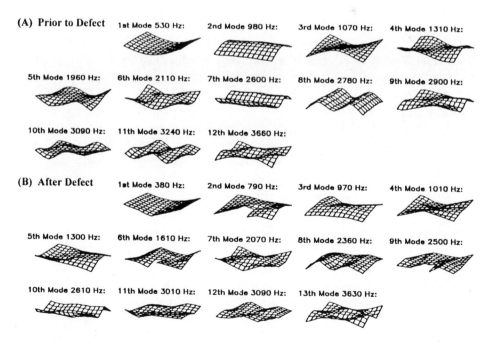

Figure 6.45. Mode shapes observed for the test specimen prior to and following the introduction of a defect [cut] (after Sun, 1991).

block modification, and frequency down-shift was observed. No significant change could be determined for the damping coefficients (z-values).

6.12.3 Modal acoustic emission

According to Gorman (1991a,b) modal acoustic emission is the study of the ultrasonic wave modes produced by acoustic emission sources in plates, rods, shells, and other thin-walled materials. Modal AE has a number of analogies with conventional modal analysis discussed briefly in the preceding section. However, where the latter considers the modes of vibration of a discrete specimen or structure, the former considers characteristics of specific modes of stress wave propagation. Furthermore, since the "source" in modal AE is actually some type of activated defect, the technique would be classed as a passive one.

A specimen, such as a plate, is considered thin if its thickness is less than the wavelength of the appropriate AE waves, where

$$\lambda = C_1 f \qquad \text{(Eq. 6.12)}$$

and λ is the wavelength, C_1 is the P-wave velocity and f is the AE wave frequency. For example in a typical steel, $C_1 = 5800$ m/s and f = 200 kHz, and based on Equation 6.12 $\lambda = 0.029$ m. In this case specimens or structures with a thickness less than 0.029 m would be considered "thin-walled". Generally, thin-walled structures are not common in the geotechnical area. However, a limited number of cases arise in mining and civil engineering including certain forms of concrete structures, tunnel and shaft lining, mechanical waveguides, etc. Therefore, details of the basic modal AE concept are included here.

According to Gorman (1991a):

"Improvements in AE practice will occur as the connections between theory and measurement technology strengthen. One possible route is to take advantage of the insights provided by simpler theoretical formulations such as plate wave theory. Such theory can be used to predict the velocities and shapes of AE waveforms from different sources."

Waves generated in a plate are termed "plate waves" and exhibit two modes of propagation, extensional and flexural. Due to the Poisson effect there is both "in-plane" and "out-of-plane" components. Using Modal AE technique these various propagation modes may be separated.

Gorman indicates that the fundamentals of Modal AE have been fairly well established, and that the basis may be found in the current literature (e.g. Gorman 1991a,b, 1994). A brief review of these basics are included here.

Figure 6.46 shows a simplified block diagram of a modal AE data acquisition system. In general this system is similar in form to a typical hybrid-digital

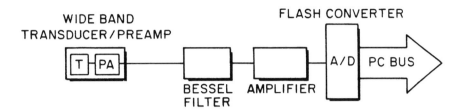

Figure 6.46. Simplified block diagram of a modal AE data acquisition system (after Gorman, 1994).

system (see Figure 3.23, Chapter 3). These are, however, considerable differences in the two systems and the following are typical requirements for a modal AE systems.

1. *Transducer*: The transducer is a flat, wide band type, similar to an accelerometer in form but displacement calibrated. Figure 6.47 illustrates characteristics of this type of transducer compared to a typical resonant-type.
2. *Filter*: Bessel-type filters are employed rather than conventional Butterworth-types to insure that associated phase shifts are linear with frequency.
3. *A/D converter*: The A/D converter is capable of high digitization rates. 12 bit vertical resolution is desirable and 12 bits at 25 MHz is suitable for most studies.
4. *Throughput rate*: Waveform capture and storage is an extremely important aspect of a modal AE system. Rates of 2 Mb/s directly to hard drive are desirable.

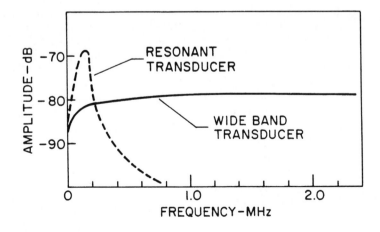

Figure 6.47. Simplified transducer response curves for wide band and resonant transducers.

Once the AE/MS signals have been captured, sophisticated signal processing techniques, developed in the electrical engineering and acoustics fields, are utilized to investigate the behavior of specific traveling wave modes. For example, plate waves are highly dispersive, changing shape as they propagate. Dispersive characteristics of relevant modes may be predicted by theory and programmed into the signal analysis facility. Furthermore, application of the cross-correlation function and the Hilbert transform may be used to determine more accurate arrival-time-difference data, allowing improved source location analysis which is independent of both threshold and gain settings. A factor of particular importance using modal AE is the greatly improved ability to identify sources and the enhanced ability to separate signals from noise. As emphasized by Gorman (1994), "this should not be taken lightly: the difficulty in separating true AE signals from noise, especially mechanical noise from frictional rubbing or fretting, pump noise, flow noise, and environmental noise has long plagued the technology".

As the availability of suitable commercial equipment, and experienced researchers increases, modal AE technique will find increasing application in the study of geologic structures and materials.

REFERENCES

Albright, J. and Pearson, C. 1984. Microseismic Activity Observed During Depressurization of an Oil Storage Cavern in Rock Salt, *Proceedings Third Conference on Acoustic Emission/ Microseismic Activity in Geologic Structures and Materials*, Pennsylvania State University, October 1981, Trans Tech Publications, Clausthal-Zellerfeld, Germany, pp. 199-210.

Anon. 1969. *Log Interpretation Principles, Vol. 1*, Schulmberger Ltd.

Anon. 1981. In-Situ Sensitivity Calibration of Acoustic Emission Systems, *Technical Review*, No. 2, Appendix C, Bruel and Kjaer, Denmark pp. 38-42.

Anon. 1992. Standard Terminology for Nondestructive Examination, Designation 1316-92, *ASTM 1992 Annual Book of Standards, Vol. 03*, pp. 595-599.

Anon. 1996a. Standard Practice for Characterizing Acoustic Emission Instrumentation, ASTM Designation E750-93, *1996 Annual Book of ASTM Standards, Vol. 03.03*, Nondestructive Testing, ASTM, Conshohoken, Pennsylvania, pp. 339-347.

Anon. 1996b. Standard Guide for Determining the Reproducibility of Acoustic Emission Sensor Response, ASTM Designation E 976-94, *1996 Annual Book of ASTM Standards, Vol. 03.03*, Nondestructive Testing, ASTM, Conshohoken Pennsylvania, pp. 380-385.

Anon. 1996c. Standard Method for Primary Calibration of Acoustic Emission Sensors, ASTM Designation E1106-92, *1996 Annual Book of Standards, Vol. 03.03*, Nondestructive Testing, ASTM, Conshohoken, Pennsylvania, pp. 491-500.

Anon. 1996d. Standard Guide for Acousto-Ultrasonic Assessment of Mechanical Properties of Composites, Laminates and Bonded Joints, *1996 ASTM Standards, Vol. 03.03*, E1495-94, Nondestructure Testing, ASTM, West Conshohocken, Pennsylvania, pp. 768-775.

Atkinson, B.K., MacDonald D. and Meredith, P.G. 1984. Acoustic Response of Fracture Mechanics of Granite Subjected to Thermal and Stress Cycling Experiments, *Proceedings of Third Conference on Acoustic Emission/Microseismic Activity in Geologic Structures and Materials*, Trans Tech Publications, Clausthal-Zellerfeld, Germany, pp. 5-18.

Baria, R., Hearn, K. and Batchelor, A.S. 1989. Induced Seismicity During the Hydraulic Stimulation of a Potential Hot Dry Rock Geothermal Reservoir, *Proceedings Fourth Conference on Acoustic Emission/ Microseismic Activity in Geologic Structures and Materials*, Pennsylvania State University, October 1985, Trans Tech Publications, Clausthal-Zellerfeld, Germany, pp. 327-352.

Barr, S.P. 1993. *The Kaiser Effect of Acoustic Emissions for the Determination of In-Situ Stress in the Carnmenellis Granite*, Ph.D. Dissertation, Camborne School of Mines, University of Exeter, United Kingdom, August 1993.

Beall, F.C. & Vary, A. (eds.), 1994. Special Issue–Acousto-Ultrasonics, *Journal Acoustic Emission, Vol. 12*, No. 1/2.

Bow, S. 1992. *Pattern Recognition and Image Preprocessing*, Marcel Dekker, Inc. New York.

Breckenridge, F.R., Proctor, T.M., Hsu, N.N., Fick, S.E. and Eitzen, D.G. 1990. Transient Sources for Acoustic Emission Work, *Progress in Acoustic Emission V*, Proceedings 10th International Symposium, Sendai, Japan, October 1990, pp. 20-37.

Brune, J.N. 1970. Tectonic Stress and the Spectra of Shear Waves from Earthquakes, *J. Geophysical Research, Vol. 75*, pp. 4997-5009.

Brune, J.N. 1971. Correction to Brune 1970, *J. Geophysical Research, Vol. 76*, p. 5002.

Cerni, R.H. and Foster, L.E. 1962. *Instrumentation for Engineering Measurement*, John Wiley and Sons, Inc., New York, pp. 1-51.

Chen, C.H. 1973. *Statistical Pattern Recognition*, Hayden Book Co., Rochelle Park, N.J.

Chkouratnik, V.L. and Lavrov, A.V. 1997. Numerical 2D-Simulation of Memory Effects in Rocks Around a Borehole, *Rock Stress*, K. Sugawa and Y. Obara (eds.), A. A. Balkema, Rotterdam, pp. 193-196.

Cranswick, E. and Sembera, E. 1988. Earthquake Site/Source Studies in the AE/MS Domain, *Proceedings Fourth Conference on Acoustic Emission/Microseismic Activity in Geologic Structures and Materials*, The Pennsylvania State University, October 1985, Trans Tech Publications, Clausthal-Zellerfeld, Germany, pp. 375-402.

Dayhof, J. 1990. *Neural Network Architectures-An Introduction,* Van Nostrand Reinhold, New York.

Diederichs, U., Schneider U. and Terrien, M. 1983. Fracture Mechanics of Concrete-Developments in Civil Engineering, Elsevier, Amsterdam, Vol. 7, pp. 157-205.

Dobrin, M. B. 1960. *Introduction to Geophysical Prospecting*, McGraw-Hill Book Company, New York, 446 pp.

Doebelin, E.O. 1975. *Measurement Systems - Applications and Design*, McGraw-Hill Book Company, 772 pp.

Dresen, L. & Ruter, H. 1994. *Seismic Coal Exploration, Part B: In-Seam Seismics*, Pergamon, Elsevier Science Inc., New York, 433 pp.

Drouillard, T.F. and Vary, A. 1994. AE Literature - Acoustic-Ultrasonic Reflections, *Journal Acoustic Emission, Vol. 12*, No. 1/2, pp. 71-78.

Dunegan, H.L. 1982. *Design and Use of Waveguides*, Acoustic Emission Instruments and Testing, Dunegan/Endevco, San Juan Capistrano.

Dunegan, H.L. 1991. Acoustic Emission Testing of Structures by Use of Mechanical Waveguides, *4th World Meeting on Acoustic Emission and 1st International Conference on Acoustic Emission in Manufacturing*, Boston, September 1991, American Society Nondestructive Testing, S.J. Vahaviolas (ed.), pp. 110-117.

Dürr, K. and Meister, D. 1984. Evaluation of Pillar Deformation and Stability in a Salt Mine, Utilizing Acoustic Emission, Mine Survey and Rock Deformation Data, *Proceedings Third Conference on Acoustic Emission/Microseismic Activity in Geologic Structures and Materials*, The Pennsylvania State University, October 1981, Trans Tech Publications, Clausthal-Zellerfeld, Germany, pp. 283-302.

Dziewonski, A.M., Chou, T.A. and Woodhouse, J.H. 1981. Determination of Earthquake Source Parameters from Waveform Data for Studies of Global and Regional Seismicity, *Journal Geophysics Research, Vol. 86*, pp. 2825-2852.

Ebel, J.E. and Bonjer, K.P. 1990. Moment Tensor Inversion of Small Earthquakes in Southwestern Germany for the Fault Plane Solution, *Geophysics Journal International, Vol. 101*, pp. 133-146.

Ewins, D.J. 1986. *Modal Testing: Theory and Practice*, Research Studies Press Ltd., Herts, England.

Fail, J.P., Grau, G. and Lavergne, M. 1962. Coupling Seismographics with the Ground, [In French], *Geophysical Prospecting, Vol. 10*, No. 2, pp. 128-147.

Feknous, N., Ballivy, G and Piasta, Z. 1995. Identification of Stress Levels in Rock with Acoustic Emission Assisted by Pattern Recognition Techniques, *Proceedings Fifth Conference on Acoustic Emission/Microseismic Activity in Geologic Structures and Materials*, The Pennsylvania State University, June 1991, Trans Tech Publications, Clausthal-Zellerfeld, Germany, pp. 135-149.

Frantii, G.E. 1977. Seismissions and Surface Waves Related to Geologic Structures, *Proceedings First Conference on Acoustic Emission/Microseismic Activity in Geologic Structures and Materials*, The Pennsylvania State University, June 1975, Trans Tech Publications, Clausthal-Zellerfeld, Germany, pp. 271-289.

Friedel, M.J. and Thill, R.E. 1990. *Stress Determination in Rock Using the Kaiser Effect*, RI 9286, U.S. Bureau of Mines.

Fu, K.S. 1982. *Applications of Pattern Recognition*, CRC Press, Inc., Boca Raton, Florida.

Gibowicz, S.J. 1993. Seismic Moment Tensor and the Mechanism of Seismic Events in Mines, *Rockbursts and Seismicity in Mines*, P. Young (ed.), Balkema, Rotterdam, pp. 149-155.

Gibowicz, S.J. and Kijko, A. 1994. *An Introduction to Mining Seismology*, Academic Press, New York, pp. 176-223.

Gorman, M.R. 1991a. Acoustic Emission for the 1990s, *Proceedings 1991 Ultrasonics Symposium*, IEEE, New York, pp. 1039-1046.

Gorman, M.R. 1991b. Plate Wave Acoustic Emission, *Journal Acoustical Society of America, Vol. 90*, No. 1, pp. 358-364.

Gorman, M.R. 1994. New Technology for Wave Based Acoustic Emission and Acousto-Ultrasonics, *AMD, Vol. 188*, Wave Propagation and Emerging Technologies, ASME, pp. 47-59.

Grabec, I. 1992. Prediction of Chaotic AE Signals by a Neural Network, *Progress in Acoustic Emission VI*, Proceedings 11th International Acoustic Emission Symposium, Fukuoka 1992, T. Kishi, K. Takahashi and M. Ohtsu (eds.), Japanese Society for Non-Destructive Inspection, Tokyo, pp. 17-24.

Grabec, I., Sachse, W. and Govekar, E. 1991. Solving AE Problems by a Neural Network, *Acoustic Emission: Current Practice and Future Directions*, ASTM, STP 1077, ASTM, West Conshohocken, pp. 165-182.

Greenfield, R.J. 1977. Amplitudes and Spectra from Underground Sources, *Proceedings First Conference on Acoustic Emission/ Microseismic Activity in Geologic Structures and Materials*, Pennsylvania State University, June 1975, Trans Tech Publications, Clausthal-Zellerfeld, Germany, pp. 405-425.

Gutenberg, B. & Richter, C.R. 1949. *Seismicity of the Earth*, Princeton University Press.

Hardy, H.R. Jr. 1974. *Microseismic Techniques Applied to Coal Mine Safety*, Final Report U.S. Bureau of Mines, Grant No. G0101743 (MIN-45), USBM Open File #23-75, NTIS-PB 241-2081AS, 208 pp.

Hardy, H.R. Jr. 1985. *Some Thoughts on the Use of Mechanical Waveguides in AE/MS Geotechnical Applications*, Internal Report RML-IR/85-5, Geomechanics Section, Mineral Engineering Department, The Pennsylvania State University.

Hardy, H.R. Jr. 1986. Source Location Velocity Models for AE/MS Field Studies in Geologic Materials, *Proceedings 8th International Acoustic Emission Symposium*, Tokyo, Japan, October 1986, Japanese Society for Nondestructive Inspection, pp. 365-388.

Hardy, H.R. Jr. 1992. Laboratory Studies Relative to the Development of Mechanical Waveguides for Acoustic Emission Monitoring of Geologic Structures, *Italian Journal Nondestructive Testing and Diagnostics, Vol. XIII*, No. 2, pp. 32-38.

Hardy, H.R. Jr. 1998. Acoustic Emission in Salt During Elastic and Inelastic Deformation, *Proceedings, Sixth Conference on Acoustic Emission/Microseismic Activity in Geologic Structures and Materials*, The Pennsylvania State University, June, 1996, Trans Tech Publications, Clausthal-Zellerfeld, Germany, pp. 15-27.

Hardy, H.R. Jr. and Beck, L.A. 1977. *Microseismic Monitoring of a Longwall Coal Mine: Volume 2 - Determination of Seismic Velocity*, Final Report U.S. Bureau of Mines, Grant G0144013, 230 pp.

Hardy, H.R. Jr., Belesky, R., Kimble, E.J., Mrugala, M., Hager, M.E. and Taioli, F. 1988. *A Study to Investigate the Potential of the Acoustic Emission/Microseismic Technique as a Means of Evaluating Slope Stability,*" PennDOT and U.S. Dept. of Transportation.

Hardy, H.R. Jr., Belesky, R., Mrugala, M., Kimble, E. and Hager, M. 1986. *A Study to Monitor Microseismic Activity to Detect Sinkholes*, Federal Aviation Administration, U.S. Dept. of Transportation, DOT/FAA/PM-86.34, National Technical Information Service (NTIS).

Hardy, H.R. Jr. and Mowrey, G.L. 1977. Study of Underground Structural Stability Using Near-Surface and Down-Hole Microseismic Techniques, *Proceedings International Symposium Field Measurements in Rock Mechanics*, Zurich, Vol. 1, pp. 75-92.

Hardy, H.R. Jr. and Mowrey, G.L. 1981. *A Microseismic Study of an Underground Natural Gas Storage Reservoir, Volume II - Field Data, Analysis and Results*, American Gas Association, Inc., Arlington, Virginia, A.G.A., Cat. No. L-51397.

Hardy, H.R. Jr., Mowrey, G.L. and Kimble, E.J. Jr. 1981. *A Microseismic Study of an Underground Natural Gas Storage Reservoir, Volume I - Instrumentation and Data Analysis Techniques, and Field Site Details*, American Gas Association, Inc., Arlington, Virginia, A.G.A. Cat. No. L-51396.

Hardy, H.R. Jr., Mrugala, M. and Armstrong, B.H. 1989. Investigation of Stress/ Strain Thresholds for the Generation of Acoustic Emission/Microseismic Activity in Geologic Materials, *Proceedings 12th World Conference on Non-Destructive Testing*, Amsterdam, 1989, Elsevier Science Publishers B. V., Amsterdam, The Netherlands, Vol. 1, pp. 471-477.

Hardy, H.R. Jr. and Oh, E. 1998. Calibration of Low-Frequency Acoustic Emission Transducers, *Progress in Acoustic Emission IX*, Proceedings 14th International Acoustic Emission Symposium, Hawaii, August 1998, M. A. Hamstad, T. Kishi, and K. Ono (eds.), Japanese Society for Non-Destructive Inspection, Tokyo, pp. IV/74-IV/85.

Hardy, H.R. Jr., Zhang, D. and Zelanko, J.C. 1989. Recent Studies of the Kaiser Effect in Geologic materials, *Proceedings, Fourth Conference on Acoustic Emission/Microseismic Activity in Geologic Structures and Materials*, Trans Tech Publications, Clausthal-Zellerfeld, Germany, pp. 27-55.

Hardy, H.R. Jr. et al. 1986. *A Study to Monitor Microseismic Activity to Detect Sinkholes*, Final Report to U.S. Dept. of Transportation/PA Dept. of Transportation, Project 84-32, Contract No. DOT-DTFA01-84-C-0005, 269 pp.

Hasbrouck, W.P. and Hadsell, F.A. 1976. Geophysical Exploration Techniques Applied to Western United States Coal Deposits, *Proceedings First International Coal Symposium*, London, U. K., Miller Freeman Publishers, San Francisco, USA, Editor: W.L.G. Muir, Vol. 1, pp. 256-287.

Hay, D.R. and Chan, R.W.Y. 1994. Theory of Operations, *ICEPAK 6.0 User's Manual*, Tektrend International Inc., Montreal.

Hay, D.R. and Chan, R.W.Y. 1995. Use of Artificial Intelligence Methods in the Analysis of Microseismic Data, *Proceedings Fifth Conference on Acoustic Emission/Microseismic Activity in Geologic Structures and Materials*, The Pennsylvania State University, June 1991, Trans Tech Publications, Clausthal-Zellerfeld, Germany, pp.589-596.

Hilburn, J.L. and Johnson, D.E. 1983. *Manual of Active Filter Design, 2nd Edition*, McGraw-Hill Inc., New York.

Holcomb, D.J. and Martin, R.J. III, 1985. Determining Peak Stress History Using Acoustic Emissions, *Proceedings, 25th U.S. Symposium on Rock Mechanics*, Rapid City, South Dakota, June 1985, A. A. Balkema, Boston, pp. 715-722.

Holub, K. 1998. Changes in b-Values During Coal Mining, *Proceedings Sixth Conference on Acoustic Emission/Microseismic Activity in Geologic Structures and Materials*, The Pennsylvania State university, June 1996, Trans Tech Publications, Clausthal-Zellerfeld, Germany, pp. 309-323.

Hsu, N.N. and Yamaguchi, K. 1990. A Rational Approach to Acoustic Emission Signal Analysis and System Calibration, *Progress in Acoustic Emission V*, Proceedings 10th International Acoustic Emission Symposium, Sendai, Japan, October 1990, pp. 361-368.

Hsu, N.N., Simmons, J.A. and Hardy, S.C. 1977. An Approach to Acoustic Emission Signal Analysis-Theory and Experiment, *Materials Evaluation, Vol. 35*, No. 10, pp. 100-106.

Hughson, D.R. and Crawford, A.M. 1986. Kaiser Effect Gauging: A New Method for Determining the Pre-existing In-Situ Stress from an Extracted Core by Acoustic Emissions, *Proceedings, International Symposium on Rock Stress and Rock Stress Measurements*, Stockholm, September 1986, pp. 359-368.

Ida, M., Yamaguchi, K. and Oyaizu, H. 1991. A Neural Network Application to AE Signal Processing, *Proceedings 4th World Meeting on Acoustic Emission and 1st International Conference on Acoustic Emission in Manufacturing*, ASNT, Boston 1991, pp. 519-526.

Ishimoto, M. and Iida, K. 1939. Observations sur les Seismes Enregistres par le Microsismographe Construit Dermerement, *Bull. Earthquake Research Institute, Vol. 17*, Tokyo University, pp. 443-478.

Kaiser, J. 1953. Erketnisse and Folgerungen aus der Messung von Gerauschen bei Zugbeanspruchung von Metalli en erkstoffen, *Archiv Fur das Eisenhuttenwesen, Vol. 24*, pp. 43-45.

Kanagawa, T., Hayashi, M. and Kitahara, Y. 1981. Acoustic Emission and Over-Coring Methods for Measuring Tectonic Stresses, *Proceedings, International Symposium on Weak Rocks*, Tokyo, pp. 1205-1210.

Keledy, F.C. 1975. Principles of Discrimination in Acoustic Emission Measurements, Technical Note – Trodyne Corporation, Teterboro, New Jersey.

Kimble, E.J. Jr. 1983. *Design of Notch and Bandpass Filters for AE/MS Field Studies*, Internal Report RML-IR/83-8, Geomechanics Section, Department of Mineral Engineering, The Pennsylvania State University.

Kimble, E.J. Jr. 1989. Development of the Mark II Mobile Microseismic Monitoring Facility, *Proceedings Fourth Conference on Acoustic Emission/Microseismic Activity in Geologic Structures and Materials*, The Pennsylvania State University, October 1985, Trans Tech Publications, Clausthal-Zellerfeld, Germany, pp. 431-450.

Knill, J.L., Franklin, J.A. and Malone, A.W. 1968. A Study of Acoustic Emission from Stressed Rock, *Int. J. Mech. Min., Sci., Vol. 5*, pp. 87-121.

Koerner, R.M., Lord, A.E. Jr. and McCabe, W.M. 1977. Acoustic Emission Studies of Soil Masses in the Laboratory and Field, Proceedings First Conference on Acoustic Emission/Microseismic Activity in Geologic Structures and Materials, The Pennsylvania State University, Trans Tech Publications, Clausthal-Zellerfeld, Germany, pp. 243-256.

Kojima, T. and Matsuki, K. 1990. A Fundamental Study of the Kaiser Effect in Rock for Tectonic Stress Measurement, *Proceedings, 10th International Acoustic Emission Symposium*, Sendai, Japan, October 1990, Japanese Society for Non-Destructive Inspection, pp. 468-475.

Lamer, A. 1970. Coupling Geophones to Soil, [In French], *Geophysical Prospecting Vol. 18*, No. 2, pp. 300-319.

Lavrov, A.V. 1997. Three-Dimensional Simulation of Memory Effects in Rock Samples, *Rock Stress*, K. Sugawa and Y. Obara (eds.), A. A. Balkema, Rotterdam, pp. 197-202.

Li, C. 1995. Micromechanics Modelling for Stress-Strain Behaviour of Brittle Rocks, *Int. J. for Numerical and Analytical Methods in Geomechanics, Vol. 19*, pp. 331-344.

Li, C. 1998. A Theory for the Kaiser Effect in Rock and It's Potential Applications, *Proceedings, Sixth Conference on Acoustic Emission/Microseismic Activity in Geologic Structures and Materials*, The Pennsylvania State University, June, 1996, Trans Tech Publications, Clausthal-Zellerfeld, Germany, pp. 171-185.

Li, C. and Nordlund, E. 1993a. Deformation of Brittle Rocks Under Compression with Particular Reference to Microcracks, *Mechanics of Materials, Vol. 14*, pp. 223-239.

Li, C. and Nordlund, E. 1993b. Experimental Verification of the Kaiser Effect in Rocks, *Rock Mech. and Rock Eng., Vol. 26*, pp. 333-351.

Lord, A. E. Jr., Fisk, C.L. and Koerner, R.M. 1982. Utilization of Steel Rods as Waveguides, Technical Note, *Journal Geotechnical Engineering Division, Vol. 108*, American Society of Civil Engineers, No. GT2, pp. 300-305.

Mae, I. and Nakao, N. 1968. Characteristics in the Generation of Microseismic Noises in Rocks Under Uniaxial Compressive Load, *Journal Society Materials Science, Japan, Vol. 17*, pp. 62-67.

Majer, E.L., King, M.S. and McEvilly, T.V. 1984. The Application of Modern Seismologic Methods to Acoustic Emission Studies in a Rock Mass Subject to Heating,

Proceedings Third Conference on Acoustic Emission/ Microseismic Activity in Geologic Structures and Materials, Pennsylvania State University, October 1981, Trans Tech Publications, Clausthal-Zellerfeld, Germany, pp. 499-516.

Manthei, G., Eisenblatter, J. and Salzer, K. 1998. Acoustic Emission Studies on Thermally and Mechanically Induced Cracking in Salt Rock, *Proceedings Sixth Conference on Acoustic Emission/Microseismic Activity in Geologic Structures and Materials*, The Pennsylvania State University, June 1996, Trans Tech Publications, Clausthal-Zellerfeld, Germany, pp. 245-265.

McConnell, K.G. 1995. *Vibration Testing – Theory and Practice*, John Wiley & Sons, Inc., New York, pp. 267-271.

McElroy, J.J. Jr. 1985. *In-Situ Stress Determination in Rock Using the Acoustic Emission Techniques*, Ph.D. Thesis, Drexel University, Philadelphia, Pennsylvania, 349 pp.

Meister, D., Spies, T., Hensel, G., Hente, B., Quijano, A. and Wallmuller, R. 1998. "Acoustic Emission and Rock Deformation Measurements as Tools for Evaluating Pillar Stability in the Asse Salt Mine," *Proceedings Sixth Conference on Acoustic Emission/Microseismic Activity in Geologic Structures and Materials*, The Pennsylvania State University, June 1996, Trans Tech Publications, Clausthal-Zellerfeld, Germany, pp. 269-296.

Michihiro, K., Yohioka, H., Hata, K. and Fujiwara, T. 1989. Strain Dependence of the Kaiser Effect on Various Rocks, *Proceedings Fourth Conference on Acoustic Emission/Microseismic Activity in Geologic Structures and Materials*, Trans Tech publications, Clausthal-Zellerfeld, Germany, pp. 87-95.

Miller, R.K. 1987. Acoustic Emission Sensors and their Calibration, *Nondestructive Testing Handbook, (Second Edition)*, Vol. 5, Acoustic Emission Testing, American Society for Nondestructive Testing, Inc., Columbus, Ohio, pp. 122-134.

Mlaker, P. F., Walker, R.E., Sullivan, B.R., Chiarito, V.P. and Malhotra, V.M. 1984. *In-Situ Nondestructive Testing of Concrete*, American Concrete Institute, SP-92, pp. 619-937.

Mogi, K. 1962. Magnitude Frequency Relation for Elastic Shocks Accompanying Fractures of Various Materials and Some Related Problems in Earthquakes, *Bull. Earthquake Research Institute , Vol. 40*, pp. 831-853.

Momayez, M. and Hassani, F. 1998. A Study Into the Effect of Confining Stress on Kaiser Effect, *Proceedings, Sixth Conference on Acoustic Emission/Microseismic Activity in Geologic Structures and Materials*, The Pennsylvania State University, June, 1996, Trans Tech Publications, Clausthal-Zellerfeld, Germany, pp. 188-194.

Momayez, M., Hassani, F.P. and Hardy, H.R. Jr. 1990. Stress Memory Measurement in Geological Materials Using thee Kaiser Effect of Acoustic Emission, *Proceedings, 10th International Acoustic Emission Symposium*, Sendai, Japan, October 1990, Japanese Society for Nondestructive Inspection, Tokyo, pp. 476-483.

Montoto, M. and Hardy, H.R. Jr. 1991. Kaiser Effect in Intact Rock: Current Status as a Means of Evaluating Thermal and Mechanical Loading, *Proceedings, Seventh International congress on Rock Mechanics*, Aachen 1991, A. A. Balkema, Rotterdam, pp. 569-572.

Montoto, M., Hardy, H.R. Jr., Fernández-Merayo, N. and Suárez del Rio, L.M. 1998. Microfractographic Evaluation of Granitic Rock Cores during Stress Relieve from Deep Boreholes: An AE/MS Evaluation Review, *Proceedings Sixth Conference on Acoustic Emission/Microseismic Activity in Geologic Structures and Materials*, The Pennsylvania State University, June, 1996, Trans Tech Publications, Clausthal-Zellerfeld, Germany, pp. 73-84.

Montoto, M., Ruiz de Argandofia, V.G., Calleja, L. and Suráez del Rio, L.M. 1989. "Kaiser Effect in Thermo-cycled Rocks," *Proceedings Fourth Conference on Acoustic Emission/Microseismic Activity in Geologic Structures and Materials,* Trans Tech Publications, Clausthal-Zellerfeld, Germany, pp. 97-116.

Morrison, R. 1967. *Grounding and Shielding Techniques in Instrumentation,* John Wiley and Sons, Inc., New York.

Mottahead, P. and Vance, J.B. 1989. A/E Determination of Salt Behavior Under Stress, *Proceedings Fourth Conference on Acoustic Emission/Microseismic Activity in Geologic Structures and Materials,* The Pennsylvania State University, October 1985, Trans Tech Publications, Clausthal-Zellerfeld, Germany, pp. 465-474.

Mukherjee, C., Anireddy, H.R. and Ghosh, A.K. 1998. Kaiser Effect Studies on Coking Coal Samples for Delineating the Yield Zone Around a Longwall Extraction Perimeter, *Proceedings Sixth Conference on Acoustic Emission/Microseismic Activity in Geologic Structures and Materials,* The Pennsylvania State University, June, 1996, Trans Tech Publications, Clausthal-Zellerfeld, Germany, pp. 195-209.

Murthy, C.R.L., Dattaguru, B. and Rao, A.K. 1987. Application of Pattern Recognition Concepts to Acoustic Emission Signal Analysis, *Journal Acoustic Emission, Vol. 6,* No. 1, pp. 19-28.

Nagashima, S., Moriya, H. and Niitsuma, H. 1992. Development and Calibration of Downhole Triaxial AE Detectors for Subsurface and Civil Engineering AE Measurements, *Progress in Acoustic Emission VI,* Proceedings 11th International Acoustic Emission Symposium, Fukuoka, Japan, October 1992, Japanese Society for Non-Destructive Inspection, Tokyo, pp. 407-414.

Nakajima, I. 1988. The Observation of Landslide by the Acoustic Emission Monitoring Rod, *Proceedings 9th International Acoustic Emission Symposium,* Kobe, Japan, November 1988, Japanese Society Non-Destructive Inspection, Tokyo, pp. 273-281.

Nakajima, I., Negishi, M., Ujihira, M. and Tanabe, T. 1995. Application of the Acoustic Emission Monitoring Rod to Landslide Measurement, *Proceedings Fifth Conference Acoustic Emission/Microseismic Activity in Geological Structures and Materials,* The Pennsylvania State University, June 1991, Trans Tech Publications, Clausthal-Zellerfeld, Germany, pp. 505-519.

Nakamura, Y. 1974. Spatial Filtration in AE Detection, *Second Japanese Acoustic Emission Symposium,* September 1974, Tokyo, Japan.

Oh, E. 1996. *Geophone Characterization for Improved Geologic Structure Evaluation,* Ph.D. Dissertation, Mineral Engineering Department, The Pennsylvania State University, University Park, Pennsylvania.

Oh, E. 1998. A Computer-Based Calibration Facility for Low-Frequency Acoustic Emission Transducers, *Proceedings Sixth Conference on Acoustic Emission/ Microseismic Activity in Geologic Structures and Materials,* Pennsylvania State University, June 1996, Trans Tech Publications, Clausthal-Zellerfeld, Germany, pp. 471-488.

Ohtsu, M. 1991. Simplified Moment Tensor Analysis and Unified Decomposition of Acoustic Emission Source: Application to in Situ Hydrofracturing Test, *Journal Geophysical Research, Vol. 96,* pp. 6211-6221.

Ohtsu, M. 1995. Acoustic Emission Theory for Moment Tensor Analysis, *Research Nondestructive Evaluation, Vol. 6,* pp. 169-184.

Ohtsu, M. and Ono, K. 1987. Pattern Recognition Analysis of Acoustic Emission from Unidirectional Carbon Fibre-Epoxy Composites Using Autoregressive Modeling, *Journal Acoustic Emission, Vol. 6,* No. 1, pp. 61-71.

Ohtsu, M., Kunitaka, A. and Yuyama, S. 1994. Post-Analysis of SIGMA Solutions for Error Estimation in Reinforced Concrete Members, *Progress in Acoustic Emission VII*, Proceedings 12th International Acoustic Emission Symposium, Sapporo 1994, T. Kishi, Y. Mori and M. Enoki (eds.), Japanese Society for Non-Destructive Inspection, Tokyo, pp. 411-416.

Ono, K. and Huang, Q. 1994. Pattern Recognition Analysis of Acoustic Emission Signals, *Progress in Acoustic Emission VII*, Proceedings 12th International Acoustic Emission Symposium, Sapporo 1994, Editors: T. Kishi, Y. Mori and M. Enoki, Japanese Society for Non-Destructive Inspection, Tokyo, pp. 69-78.

Pollock, A. 1979. *Be More Discriminating*, Technical Note - Dunegan/Endevco, San Juan Capistrano, California.

Power, D.V. 1977. Acoustic Emissions Following Hydraulic Fracturing in a Gas Well, *Proceedings First Conference on Acoustic Emission/ Microseismic Activity in Geologic Structures and Materials*, Pennsylvania State University, June 1975, Trans Tech Publications, Clausthal-Zellerfeld, Germany, pp. 291-308.

Redwood, M. 1960. *Mechanical Waveguides*, Pergamon Press, Inc., New York.

Rothman, R.L. 1977. Acoustic Emission in Rock Stressed to Failure, *Proceedings, First Conference on Acoustic Emission/Microseismic Activity in Geologic Structures and Materials*, The Pennsylvania State University, June 1975, Trans Tech Publications, Clausthal-Zellerfeld, Germany, pp. 109-133.

Scholz, C.H. 1968. The Frequency-Magnitude Relation of Microfracturing in Rock and its Relation to Earthquakes, *Bull. Seismologic Society America, Vol. 58*, No. 1., pp. 399-415.

Seto, M., Vutukuri, V.S. and Nag, D.K. 1997. Possibility of Estimating In-Situ Stress of Virgin coal Field Using Acoustic Emission Technique, *Rock Stress*, K. Sugawa and Y. Obara (eds.), A. A. Balkema, Rotterdam, pp. 463-468.

Shen, H. W. 1995. Objective Kaiser Stress Evaluation in Rock, *Proceedings Fifth Conference on Acoustic Emission/Microseismic Activity in Geologic Structures and Materials*, The Pennsylvania State University, June 1991, Trans Tech Publications, Clausthal-Zellerfeld, Germany, pp. 177-196.

Shen, W.H. 1996. *Acoustic Emission Potential for Monitoring Cutting and Breakage Characteristics of Coal*, Ph.D. Dissertation, Mineral Engineering Department, The Pennsylvania State University, 249 pp.

Shen, W.H. 1998. Application of Acoustic Emission Techniques to the Monitoring of Cutting and Breakage of Geologic Materials, *Proceedings Sixth Conference on Acoustic Emission/Microseismic Activity in Geologic Structures and Materials*, The Pennsylvania State University, June 1996, Trans Tech Publications, Clausthal-Zellerfeld, Germany, pp. 85-197.

Simpson, P.K. 1989. *Artificial Neural Systems*, Pergamon Press, Oxford.

Stuart, C.E., Meredith, P.G., Murrell, S.A.F. and Van Munster, H. 1995. Influence of Aniostropic Crack Damage Development on the Kaiser Effect Under True Triaxial Stress Conditions, *Proceedings Fifth Conference on Acoustic Emission/Microseismic Activity in Geologic Structures and Materials*, The Pennsylvania State University, June 1991, Trans Tech Publications, Clausthal-Zellerfeld, Germany, pp. 205-219.

Studt, T. 1991. Neural Networks: Computer Toolbox for the 90s, *R&D Magazine*, September 1991, pp. 36-42.

Sun, X. 1991. *A Feasibility Study of the Application of Modal Analysis to Geotechnical Engineering*, Ph.D. Dissertation, Mineral Engineering Department, The Pennsylvania State University, December 1991.

Sun, X. and Hardy, H.R. Jr. 1990. A Feasibility Study of Modal Analysis in Geotechnical Engineering: Laboratory Phase, *Proceedings 31st US Rock Mechanics Symposium*, A.A. Balkema, Rotterdam, pp. 661-668.

Sun, X. and Hardy, H.R. Jr. 1992. A Feasibility Study of Modal Analysis in Geotechnical Engineering: Field Phase, *Proceedings 33rd US Rock Mechanics Symposium*, A. A. Balkema, Rotterdam, pp.1019-1028.

Sun, X., Hardy, H.R. Jr. and Rao, M.V.M.S. 1991. Acoustic Emission Monitoring and Analysis Procedures Utilized During Deformation Studies on Geologic Materials, *Acoustic Emission: Current Practice and Future Directions*, ASTM STP 1077, W. Sachse, J. Roget and K. Yamaguichi (eds.), ASTM, Philadelphia, pp. 365- 380.

Suzuki, Z. 1953a. A Statistical Study on the Occurrence of Small Earthquakes – I, *Science Reports Tohoku University, Fifth Series Geophysics, Vol. 5*, pp. 177-182.

Suzuki, Z. 1953b. A Statistical Study on the Occurrence of Small Earthquakes - II, *Science Reports Tohoku University, Fifth Series Geophysics, Vol. 6*, pp. 105-118.

Suzuki, H., Kinjo, T., Hayaski, Y., Takemato, M. and Ono, K. 1996. Wavelet Transform of Acoustic Emission Signals, *Journal Acoustic Emission, Vol. 14*, No. 2, pp. 69-84.

Talebi, S., Cornet, F.H. and Martel, L. 1989. Seismo-Acoustic Activity Generated in a Granitic Rock Mass, *Proceedings Fourth Conference on Acoustic Emission/ Microseismic Activity in Geologic Structures and Materials*, The Pennsylvania State University, October 1985, Trans Tech Publications, Clausthal-Zellerfeld, Germany, pp. 491-509.

Trombik, M. and Zuberek, W.M. 1977. Microseismic Research in Polish Coal Mines, *Proceedings First Conference on Acoustic Emission/Microseismic Activity in Geologic Structures and Materials*, The Pennsylvania State University, June 1975, Trans Tech Publications, Clausthal, Germany, pp. 169-194.

Wang, F. 1999. *Experimental Study of Kaiser Effect and AE Characteristics of Rocks Under Various Loading Conditions*, Ph.D. Dissertation, University of New South Wales, Australia, July 1999.

Washburn, H. and Wiley, H. 1941. The Effect of the Placement of a Seismometer on it's Response Characteristics, *Geophysics, Vol. 6*, pp. 116-131.

Wasserman, P.D. 1990. *Neural Computing-Theory and Practice*, Van Nostrand Reinhold, New York.

Wasserman, P.D. and Schwartz, T. 1988. Neural Networks, *IEEE Expert, Vol. 3*, Spring 1988, No. 1, pp. 10-16.

Watters, R.J. and Roberts, K. 1995. The Kaiser Effect and It's Applicaiton to Slope Instability, *Proceedings Fifth Conference on Acoustic Emission/Microseismic Activity in Geologic Structures and Materials*, The Pennsylvania State University, June 1991, Trans Tech Publications, Clausthal-Zellerfeld, Germany, pp. 233-255.

Wong, I.G., Humphrey, J.R. and Silva, W.J. 1989. Microseismicity and Subsidence Associated With a Potash Solution Mine, Southeastern Utah, USA, *Proceedings Fourth Conference on Acoustic Emission/Microseismic Activity in Geologic Structures and Materials*, The Pennsylvania State University, October 1985, Trans Tech Publications, Clausthal-Zellerfeld, Germany, pp. 287-306.

Wood, B.R.A., Flynn, T.G., Harris, R.W. and Moyes, L.M. 1990. Comparisons Between Waveguides for Three Long Term Acoustic Emission Monitoring Projects, Proceedings 10th International Acoustic Emission Symposium, Tohoku University, Sendai, Japan, October 1990, Japanese Society Nondestructive Inspection, Tokyo, pp. 501-506.

Wood, B.R.A., Harris, R.W. and Davis, S. 1995. An Evaluation of End Cap Materials, Kaiser Effect and Wave Propagation Characteristics in Rock Samples, *Proceedings, Fifth Conference on Acoustic Emission/Microseismic Activity in Geologic Structures and Materials*, The Pennsylvania State University, June 1991, Trans Tech Publications, Clausthal-Zellerfeld, Germany, 625-637.

Yoshikawa, S. and Mogi, K. 1990. Experimental Studies on the Stress History on Acoustic Emission Activity – A Possibility of Estimation of Rock Stress, *J. Acoustic Emission, Vol. 8*, No. 4, pp. 113-123.

Yuki, H. and Homma, K. 1992. AE Source Waveform Analysis by Using a Neural Network, *Progress in Acoustic Emission VI*, Proceedings 11th International Acoustic Emission Symposium, Fukuoka 1992, T. Kishi, K. Takahashi and M. Ohtsu (eds.), Japanese Society for Non-Destructive Inspection, Tokyo, pp. 235-242.

Yuyama, S., Okamato, T., Shigeishi, M. and Ohtsu, M. 1994. Quantitative Evaluation and Visualization of Cracking Process in Reinforced Concrete Specimens by a Moment Tensor Analysis of Acoustic Emission, *Progress in Acoustic Emission VII*, Proceedings 12th International Acoustic Emission Symposium, Sapparo 1994, T. Kishi, Y. Muri and M. Enoki (eds.), Japanese Society for Non-Destructive Inspection, Tokyo, pp.347-354.

Zhang, B.Q. 1989. Linear Location of AE Simulated Sources on Steel Pipelines with Waveguides, World Meeting Acoustic Emission, Charlotte, North Carolina, March 1989, Special Supplement, *Journal of Acoustic Emission, Vol. 8*, No. 1-2, pp. S49-S52.

Zhang, B.-C., Li, H., Li, F.-Q. and Shin, K. 1998. Kaiser Effect Tests on Oriented Rock Core, *Proceedings Sixth Conference on Acoustic Emission/Microseismic Activity in Geologic Structures and Materials*, The Pennsylvania State University, June 1996, Trans Tech Publications, Clausthal-Zellerfeld, Germany, pp. 212-224.

Zuberek, W.M. 1989. Application of the Gumbel Asymptotic Extreme Value Distributions to the Description of the Acoustic Emission Amplitude Distribution, *Proceedings Fourth Conference on Acoustic Emission/Microseismic Activity in Geologic Structures and Materials*, The Pennsylvania State University, October 1985, Trans Tech Publications, Clausthal-Zellerfeld, Germany, pp. 649-665.

Zuberek, W.M., Zogala, B. and Odudiel, R. 1998. Laboratory Investigations of the Maximum Temperature Memory Effect in Sandstone with Measurements of Acoustic Emission and P-Wave Velocity, *Proceedings Sixth Conference on Acoustic Emission/Microseismic Activity in Geologic Structures and Materials*, The Pennsylvania State University, June 1996, Trans Tech Publications, Clausthal-Zellerfeld, Germany, pp. 157-168.

APPENDIX

Manufacturers and Suppliers of AE/MS Transducers, Equipment and Components

1. *AE/MS Transducers*

  1.1 *Accelerometers*

Brüel & Kjaer
Spectris Technologies Inc.
2364 Park Central Blvd.
Decatur
GA 30035-3987, USA
Tel: (770) 981-3998
FAX: (770) 987-8704

Endevco
30700 Rancho Viejo Road
San Juan Capistrano
CA 92675, USA
Tel: (800) 289-9204
FAX: (714) 661-7231

Kistler Instruments Corp.
75 John Glenn Drive
Amherst
New York 14228-2171, USA
Tel: (800) 755-5746

PCB Piezotronics Inc.
3425 Walden Ave.
Depew
NY 14043, USA
Tel: (716) 684-0001
FAX: (716) 684-0987

Wilcoxon Research
1 Firstfield Road
Gaithersburg
MD 20878, USA
Tel: (301) 330-8801
FAX: (301) 330-8873

1.2   *AE-Transducers*

Digital Wave Corporation
11234-A East Caley Avenue
Englewood
CO 8011, USA
Tel: (303) 790-7559
FAX: (303) 790-7567

Industrial Quality Inc.
640 East Diamond Ave.
Gaithersburg
MD 20877, USA
Tel: (301) 948-2460
FAX: (301) 948-9037

**Dunegan Engineering
Consultants, Inc.**
P.O. Box 1749
San Juan Capistrano
CA 92623, USA
Tel: (714) 661-8105
FAX: (714) 661-3723

**Physical Acoustics
Corporation**
P.O. Box 3135
Princeton
NJ 08543-3135, USA
Tel: (609) 716-4000
FAX: (609) 716-0706

1.3   *Geophones*

Bison Instruments, Inc.
560-T, Rowland Road
Minneapolis
MN 55343, USA
Tel: (612) 931-0051
FAX: (612) 931-0997

Oyo Geospace
7334 North Gessmer St.
Houston
TX 77040, USA
Tel: (713) 666-1611

Sercel, Inc.
(Formally Mark Products, Inc.)
1052 Fallstone Road
Houston
TX 77099, USA
Tel: (281) 498-0600

**Western Geophysical
Exploration Products**
Western Atlas International
3600 Briarpark Drive
Houston
TX 77042-4299, USA
Tel: (713) 964-6093
FAX: (713) 964-6500

1.4   *Semi-Conductor Strain Gages*

BLH Electronics Inc.
75 Showmut Road
Canton
MA 02021, USA
Tel: (617) 821-2000
FAX: (617) 828-1451

1.5 *Piezo-Electric Elements*

Krautkramer
50 Industrial Park Road
Lewistown
PA 17044, USA .
Tel: (717) 242-0327

Staveley Sensors, Inc.
91 Prestige Park Circle
East Hartford
CT 06108, USA
Tel: (800) 569-1408

Polytec PI
3152 Redhill Ave.
Suite 110
Costa mesa
CA 92626, USA
Tel: (714) 850-1835
FAX: (714) 850-1831

Valpey-Fisher
75 South St.
Hopkinton
MA 01748, USA
Tel: (508) 435-6831
FAX: (508) 435-5289

2. *AE/MS Monitoring Systems*

2.1 *High Frequency Systems*

**Digital Wave
Corporation (*)**

**Dunegan Engineering
Consultants, Inc. (*)**

**Physical Acoustics
Corporation (*)**

Vallen-Systeme GmbH
c/o Acoustic Technology
Group
2644 La Via Way
Sacramento
CA 95825-0307, USA
Tel: (916) 483-1311
FAX: (916) 483-2124

2.2 *Low Frequency Systems*

Electro-Lab
Building 10
Spokane Industrial Park
Spokane
WA 99216, USA
Attention: Mr. Corwin Mallot
Tel: (509) 928-0929

---

* Address of manufacturer or supplier presented earlier in appendix

3. Miscellaneous Equipment

3.1 High Frequency Calibration System

Acoustic Emission Associates
c/o Acoustic Technology Group (*)

3.2 Low Frequency Calibration - Air Guns

Bolt Technology
Four Duke Place
Norwalk
CT 06854
Tel: (203) 853-0700

---

*Address of manufacturer or supplier presented earlier in appendix.

SUBJECT INDEX

AUTHOR INDEX